NEW TRENDS IN INDUSTRIAL AUTOMATION

Edited by **Pengzhong Li**

New Trends in Industrial Automation
http://dx.doi.org/10.5772/intechopen.75341
Edited by Pengzhong Li

Contributors

Liang Luo, Tomasz Rymarczyk, Shane Loughlin, Rahul Singh Chowhan, Pengzhong Li

First published in London, United Kingdom, 2019 by IntechOpen
IntechOpen is the global imprint of INTECHOPEN LIMITED, registered in England and Wales, registration number: 11086078, The Shard, 25th floor, 32 London Bridge Street
London, SE19SG – United Kingdom
Printed in Croatia

British Library Cataloguing-in-Publication Data
A catalogue record for this book is available from the British Library

Additional hard copies can be obtained from orders@intechopen.com

New Trends in Industrial Automation, Edited by Pengzhong Li
p. cm.
Print ISBN 978-1-78985-951-5
Online ISBN 978-1-78985-952-2

Meet the editor

Professor Dr. Pengzhong Li is the Assistant Director of Sino-German School of Postgraduate Studies (CDHK) of Tongji University. He received his Ph.D. degree in Mechanical Engineering from Tongji University in 2004. From 1995 to 2001, he worked as the Manager of Business Department and Warehousing Management Department in Guilin Daewoo Bus Co., LTD. He is a Director of Intelligent Manufacturing and Services Branch of China Creative Studies Institute and Director of East China Branch of National College Institute of Manufacturing Automation. His research interests are intelligent manufacturing systems and manufacturing process quality control technology. At present, he is committed to research on man-machine coordination in the application of Artificial Intelligence.

Contents

Preface

In the era of intelligent manufacturing, digital twin technology, industrial internet technology, image recognition, new signal processing technology, and the extreme improvement of computing performance of control modules, the technical connotation of industrial automation, control objects, and control objectives have also been greatly expanded. At the same time, system modeling technology and joint simulation technology provide new research methods and technical means for the design and optimization of industrial control systems. These new technological developments have promoted the development of industrial automation technology in the direction of intellectualization.

This book is a collection of articles on novel approaches to current problems in industrial automation by academicians, researchers, and practicing engineers. The book is divided into five chapters and encompasses multidisciplinary areas within industrial automation, such as distributed system architecture, application of non-destructive distributed sensor system, joint simulation technology, and a summary of intelligent transition. *New Trends in Industrial Automation* provides useful reference material for technical workers in the field of industrial automation, including undergraduate and postgraduate students, academicians, researchers, and practicing engineers.

Pengzhong Li
Chinesisch-Deutsches Hochschulkolleg
Tongji University
Shanghai, P.R. China

Chapter 1

Introductory Chapter: New Trends in Industrial Automation

Pengzhong Li

Additional information is available at the end of the chapter

http://dx.doi.org/10.5772/intechopen.84772

1. Introduction

As a comprehensive technology, industrial automation is a general term for information processing and process control of measurement, manipulation, etc., without direct manual intervention, according to expected aim in machine equipment or production process. Through the application of computer, electronic equipment, control theory, and related process technologies, industrial automation produces the management functions of optimization, detection, control, and regulation of the whole industrial production process to realize the established objectives, achieving industrial production increase, energy saving, consumption reduction, and safe production.

The foundation of intelligent manufacturing is digitalization, networking, and integration. Correspondingly, industrial automation in the era of intelligence will transform centralized control into decentralized enhanced control under the original automation technology and architecture, so that the communication between sensors and the Internet can be seamlessly docked, establishing a highly flexible, personalized, and digital production mode that integrates products and services. In this mode, production automation technology can make equipment more intelligent through self-diagnosis, self-correction, and various functional software to better assist workers to complete production. Therefore, the communication and integration capabilities of automation equipment are required to be stronger, while the automation software needs to have a stronger ability of analysis and processing and data sharing with other software systems of enterprises.

2. Development of modern industrial automation

With the large-scale, continuous, and highly parameterized industrial devices, the requirements for industrial automation systems are constantly increasing. In the era of intellectualization, in order to independently meet the requirements of safe start-stop, stable operation, optimal operation, and fault handling of industrial production systems, it is necessary to seamlessly integrate various instrument products and manufacturing systems into a coordinated information system. Therefore, unified control platform and smooth communication network are the basis of automation in the era of intellectualization. Industrial automation will be further expanded.

2.1. Vertical integration of automation system

Intelligence requires vertical integration of automation system with lower field sensing and data acquisition layer and upper enterprise management system. When the automation system runs, from machine operation, energy use, variable processing to material use, etc., controllers, sensors, and other equipment will produce a large amount of data in every link of the production process. The data from the production site will exceed the business data generated by the company in a few years. It is imperative to combine all the data from the plant operating systems with the information from business applications to create operational intelligence, especially remote maintenance solutions and cloud-based services, to meet the increasing demand for data analysis-based services.

2.2. Horizontal integration of intellectualization-driven automation system

Through intellectualization and its supporting technologies, the control, drive, and low-voltage distribution systems are deeply integrated to form a unified platform for automation integration, which can provide support for scalable motion and machine control in a single programming environment. This integration reduces data storage scale, while the openness of the control platform ensures easy integration with third-party components. How to deal with the data transmission, information sharing, coordinated operation, and autonomous control between these instruments and systems to meet user requirements has become a very important technology—system integration technology. In addition, the visualization and information software used on each machine needs to be standardized. At this time, the application technology aiming at the overall solutions and the application software represented by optimization software and advanced control algorithm have also become the new development trends.

2.3. Networking

2.3.1. Standard open communication network

Vertical and horizontal integration of automation systems requires a unified network infrastructure to establish a standard open communication network to realize the mutual communication between all devices in the system. In the future, network switching equipment will be more widely used. The application of independent IP enables products and devices to have identifiable independent identities, easy to track, locate, and monitor. In addition, standard communication can integrate more digital devices into the network of production line, such as cameras, RFID readers, digital tablets, safety magnetic cards, and so on, improving the refinement of production management.

2.3.2. *Networking of control*

There are two components of control system network, namely distributed control system (DCS) and fieldbus control and industrial Ethernet system (FCS). These systems clearly reflect the development direction of the current networked control system, namely, distribution, network integration, and node intelligence. In the development process, Ethernet slowly monopolizes the local area communication in the computer field, and Ethernet and fast Ethernet are gradually unified by the upper communication of process control network. The application of networked control system to Ethernet has become a trend, and the interoperability, digital interconnection, and high open network performance of Ethernet are also in line with the characteristics of fieldbus network control. Therefore, industrial Ethernet has become an important direction for the future development of fieldbus technology.

2.4. Mobile technology

Mobile technology makes management and work flexible. With access to production data and information on tablets or smartphones, factory managers and employees can "move" and communicate with production systems anytime, anywhere. In the future, cloud technology will be used to process and store data from all over the world, and real-time data will be used everywhere. People can contact any relevant personnel anytime, anywhere, and share experience and knowledge with colleagues around the world to solve business problems. No matter where the technical experts are, the call center representatives can consult them in real time, and the experts themselves can visit the history of equipment services and other devices anywhere in the world, as well as check factory updates and other consulting.

2.5. Virtualization

Virtualization reduces dependence on physical servers and other hardware while saving energy costs for factories. Virtualization technology can also improve machine reliability, create low-cost, high-availability backup solutions, and allow multiple instances of the operating system to run on a single hardware. The latest DCS systems have applied virtualized servers to achieve faster processing speed and lower life cycle costs.

3. Conclusion

In the era of intelligent manufacturing, digital twin technology, industrial Internet technology, image recognition, new signal processing technology, and the extreme improvement of computing performance of control module expand the technical connotation of industrial automation and control objects, and control objectives have also been greatly expanded. At the same time, system modeling technology and joint simulation technology provide new research methods and technical means for the design and optimization of industrial control system. These new technological developments have promoted the development of industrial automation technology to the direction of intellectualization.

Advances on studies for industrial automation have been performed by many researchers, and in the present book, some of them are presented. These advances are focused on some development trends of industrial automation to meet intelligent needs. Although the presented book does not provide a comprehensive treatment by any means to its topics, it is still a very constructive venue to direct readers' attention to some of the advanced trends of industrial automation.

Author details

Pengzhong Li

Address all correspondence to: leepz@tongji.edu.cn

Sino-German College of Postgraduate Studies, Tongji University, Shanghai, P.R. China

Chapter 2

Evolution and Paradigm Shift in Distributed System Architecture

Rahul Singh Chowhan

Additional information is available at the end of the chapter

http://dx.doi.org/10.5772/intechopen.80644

Abstract

New age programming languages highlight the mobility of objects and on-the-fly communication mechanism even being available on nodes with intermittent connections. We are breathing in the era where the working framework enables the procedures to oversee imparted information and arrangement to a domain where distinctive procedures are executing on discrete frameworks that essentially makes use of message-based correspondence or mobile communication architecture. The highlight that has been conceived for the years has spawned the remote administration and remote access in distributed computing framework and was outlined as an approach to digest the strategy call component to use between frameworks associated through a system. These frameworks contains the stub and skeleton on client and server side respectively which behaves as remote proxies and deals with marshaling and unmarshalling of the incoming and outgoing data. This has incurred the need of more distributed and platform independent communication mechanisms, that can not only make intercalling of functions but also support features like platform independency from various object oriented based programming languages. The distinctions in the programming model prompt higher state of abilities and more implicit customer side mechanisms for simple and hands-on interaction with the code that actualizes and implements the distributed frameworks.

Keywords: distributed computing, active objects, remote calling, synchronous message passing, network persistent storage

1. Introduction

The distributed system architecture field is more abstract, being at the level above the algorithms and data structures fields [1]. These architectures include the global control structures, protocols for communication, synchronization, physical distribution, scaling and performance, remote access and selection among design alternatives. At this level the common architectural styles like pipes and filters, object orientation, event-based models, and table-driven interpreter etc., and their computing techniques and distributed network architectures are seen. The level of performance within the distributed computing is directly proportional degree of multiplicity of resources involved and participating in it. This is one of the important factors that affects and regulates the usefulness within the distributed computing defining the capability of computing system to support multiplicity and migration. Most of distributed approaches involve dispersion of data over various network machines for reliability and availability of data which requires deployment of communication protocols that can server inter-platform connectivity. The interconnections in a disseminated framework allows machine to interact autonomously allowing them to share memory or processors. They can communicate with each other utilizing messages, snippets of data exchanged, intimating a function on one machine and finishing on other and so on. Simply by messages many information can be conveyed to execute with specific contentions, also they can send and get bundles of information and can do many more such things [2].

Dispersed frameworks over the network has always been communicating and imparting the information by means of message passing. This type of correspondence was exceptionally straightforward where one side (client) bundles a few information, known as a message and sends it to the opposite side where it is decoded or stored additionally. The configuration of the message and the manner by which it will be prepared by the recipient is carried out by application subordinate. In a few applications the recipient may react by sending an answer message often called as acknowledgement or response while in other cases, this won't not occur. This approach likewise makes it difficult to reuse segments of one circulated framework in other conveyed frameworks as the message is encoded in language could not be decodable or readable by other language framework with different sets of library and calling mechanisms [3]. Despite the fact that message passing can be powerful, it would be decent if there were more uniform, reusable, and easy to understand methods for getting things done remotely by calling remote function on local machine or by sending function for execution on server. Such unusualness requires an extensive variety of new procedures past those utilized as a part of conventional computing. This also involves participation of an appropriated framework by letting the compiler or run-time libraries handle various issues of scheduling and allocation.

One of the alternatives in the design of the distributed system architecture is how to access remote resources or make calls to remote objects and also how to send the program over network. Currently, the client–server paradigm is the most common style, where the code does not move at all. Under the code-mobility and remote programming domain there are certain paradigm that helps in understanding the shift happened in distributed system architecture. Like, remote method invocation allows invocation of remote objects to java enabled platforms. Code-on-demand paradigm calls upon the code from a distant site which is then downloaded and executed on the local machine. And, remote evaluation paradigm which sends the code to another site where it is executed from which result is returned back to caller [4]. These distributed mechanism helps in building up of mobile and portable code design to help effective information and program relocation in various processing states on heterogeneous execution platforms.

2. Architectural taxonomies

2.1. Centralized paradigm

Allocation of numerous resources to a small number of computers called Server-hosts, yet keeping client-hosts simpler by offloading the computation to central terminal is termed as centralized Paradigm. This type of architectural taxonomy relies heavily on network resources like servers and infrastructure for computation and storage. In this typography, client-hosts are diskless nodes that are dependent on central network terminal to load its operating system. Simply, it acts as an input/output interface to the server because they neither have their own operating system nor personalized resources. The much broader infrastructure used for such paradigm is Thin Client, which is a lightweight computer that is purposely built for remote into a server, where many client-hosts share their computations with a server or server farm. It depends heavily on another computer (server) to fulfill its computational roles. The specific roles assumed by the server-host may vary, from hosting a shared set of virtualized applications, a shared desktop stack or virtual desktop, to data processing and file storage on the behalf of client-hosts [5].

The server-side infrastructure makes use of cloud computing software such as application virtualization, hosted shared desktop (HSD) or desktop virtualization (VDI). This combination forms what is known today as a cloud based system where desktop resources are centralized into one or more data centers. Basically, this type of architecture is described by lack of delegation as they have single management station to initiate requests for low-level data.

2.2. Hierarchical paradigm

Distributed processing encompasses a wide range of task autonomy and semantic richness in hierarchical architectures. This paradigm describes implementation labels that employ vertical delegation for management functionality. Hierarchical approach includes distributed objects and limited forms of Management-by-Delegation (MbD) with code mobility technologies such as Remote EValuation (REV) and Code-on-Demand (CoD). Distributed objects describe a form of gateway operation allowing the communication with encapsulated data and actions remotely. Likewise, REV provides code for execution of intended management function while CoD retrieves and caches code to execute the intended management function [6]. The hierarchical paradigm supports the delegation as following:

a. Delegation-by-domain: Domain delegation is referred as a simplified distributed paradigm. In this, a central authority assigns complete management control of a specified domain to the domain itself. The distributed domain functions independently of the central authority. Management information is not shared, and resources and administrative control resides with the specified domain. Central authority behaves as task coordinator to delegate task to different domains.

b. Delegation by micro-task with low-level semantics: Delegation by micro task in distributed hierarchical paradigm allows the central authority to employ one or more management stations to perform specified tasks. Low-level semantics signifies the little abstraction from the details of the management task. Likewise, this method of delegation statically retrieved low-level data from simple agents before handing the response data to the central authority for processing into information.

c. Delegation by micro-task with high-level semantics: High-level semantics refers to meaningful abstractions from low-level data. For example, this method of delegation statically retrieves object data from a distributed environment before handling the response object to the central authority for processing. This framework encapsulated the protocol that supports communication between objects. Example of this distributed object paradigm includes common object request broker architecture (CORBA) and web based enterprise management (WBEM).

d. Delegation by macro-task with low-level semantics: Delegation by macro task allowed a central authority to empower one or more management stations to control specified managed elements rather than specified element properties. The management station performs necessary functions such as statically retrieving low-level data from simple agents to be processed into information by managing application. It is also responsible for taking corrective action if central authority is lost while communications.

e. Delegation by macro-task with high-level semantics: This form of delegation involves one or more authorized management stations controlling specified managed elements. Management functions include statically retrieving object data from a distributed environment which is subsequently processed by the managing application. It allows effective control decomposition and functional approximation to promote framework scalability, run time overhead reductions and workload dynamics. Example of this approach is a Goal Driven Network Management System [7].

2.3. Cooperative paradigm

Semantically rich delegation referred to a cooperative paradigm in distributed systems that empower the remote agent to control specified elements with limited instructions for preset operations. The intelligent agent relies on high-level goals and changing contextual data to make appropriate independent determination for successful management in a complex environment. Along with high autonomy and low task specification, cooperative paradigm uses horizontal delegation to cooperate with other agents unlike vertical delegation in hierarchical approaches. This is also more effective for real-time data collection within large complex and evolving networks. However, these approaches require some sort of system fidelity and measures of consistency across all nodes ensuring cooperation towards a common goal [8].

3. Client-server architecture

Distributed application structure defines client–server model that does segregation of workloads between service or resource provider, called servers and service or resource requester, called clients. These two separate components, a client and a server, which communicate over a network through a TCP/IP handshake paradigm. The client requests information, while the server responds when its advertised services are accessed. This each request/response, as depicted in **Figure 1**, is a complete round trip on the network. The code that implements these services i.e. the know-how is hosted locally by the server, also server has processing capabilities. Client decides with some intelligence which of services offered by server it should use.

3.1. One-tier architecture

Single-tier architecture is the simplest, single tier on single user, and is the equivalent of running an application on a personal computer as shown in **Figure 2**. All the components like user interface, business logic, and data storage, which are necessary to run an application, are located within the system. They are the easiest to design, but the least scalable as they are not part of a network also they cannot be used for designing web applications [9].

3.2. Two-tier architecture

Two-tier architectures supply a basic network between a client and a server. For example, the basic web model is a two-tier architecture as illustrated in **Figure 3**. A web browser makes a request from a web server, which then processes the request and returns the desired response, in this case, web pages. This approach improves scalability and divides the user interface

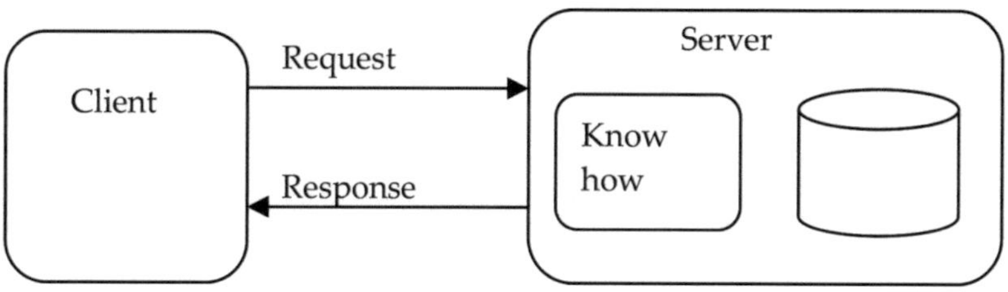

Figure 1. Client server paradigm.

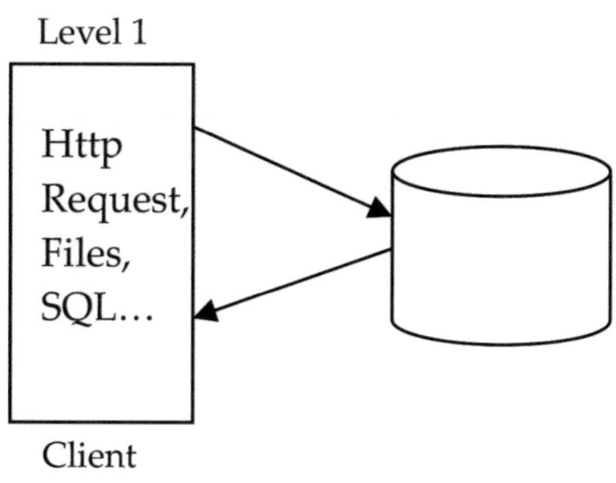

Figure 2. Single-tier architecture.

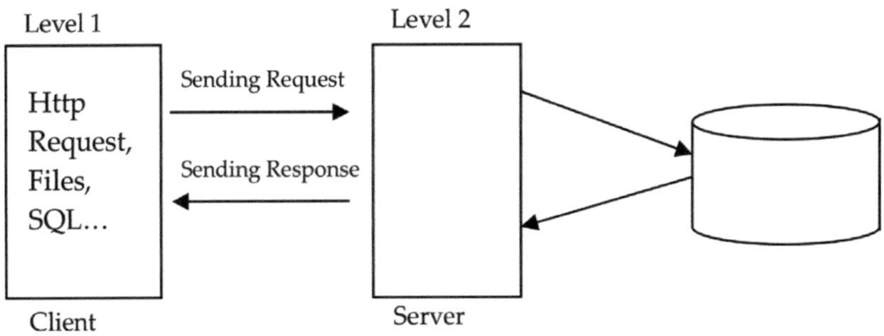

Figure 3. Two-tier architecture.

from the data layers. However, it does not divide application layers so they can be utilized separately. This makes them difficult to update and not specialized. The entire application must be updated because layers are not separated.

3.3. Three-tier architecture

Three-tier architecture is most commonly used to build web applications. In this model, the browser acts like a client, middleware or an application server contains the business logic, and database servers handle data functions. This approach separates business logic from display and data [10]. So the three layers commonly known as: presentation layer (PL/UI), business logic layer (BLL) and data access layer (DAL) as shown in **Figure 4**.

a. Presentation tier (Level 1): This provides the application's user interface (UI). Being the topmost level it displays information related to user oriented functionality responsible for managing user interaction with the system. This acts as common bridge into core business logic encapsulated in business layer.

b. Business logic tier (Level 2): This is also called application layer as it controls an application's functionality by performing detailed processing. This layer implements the core functionality of the system encapsulating the relevant business logic. It has components exposing service interfaces for callers to use.

c. Data access tier (Level 3): This includes data persistence mechanisms like database servers, file shares, etc. providing access to data hosted within system and data exposed by other networked systems. The data layer exposes generic interfaces that can be consumed by components in the business layer. It also provides an API to application layer that exposes methods of managing the stored data without out casting dependencies on the data storage mechanisms.

3.4. N-tier architecture

Terms layer and tier are often used interchangeably but one point of difference is that a layer is a logical structuring mechanism for the elements that make up the software solution. That means logical software component groups, mainly by functionality, are used for software development purpose. By contrast, a tier is a physical structuring mechanism for the system infrastructure [11].

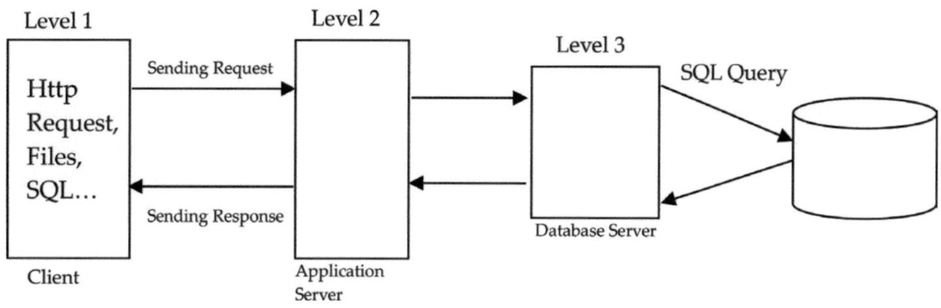

Figure 4. Three-tier architecture.

Like an individual running server is one tier and several running servers may also be counted as one tier. Layer software implementation has many advantages and is a good way to achieve N-tier architecture. Layer and tier may or may not exactly match each other. Each layer may run in an individual tier. However, multiple layers may also be able to run in one tier.

N-tier implies more than three levels or tiers involved as depicted in **Figure 5**; mostly additional tiers are associated with business logic tier. Some layers in 3-tier can be broken further into more layers. These broken layers may be able to run in more tiers. For example, application layer can be broken into business layer, persistence layer or more. Presentation layer can be broken into client layer and client presenter layer [12]. So, in order to claim a complete N-tier architecture, client presenter layer, business layer and data layer should be able to run in three separate computers (tiers).

a. Client tier: This tier is involved with users directly. There may be several different types of clients coexisting, such as WPF, Window form, HTML web page and etc.

b. Client presenter tier: This contains the presentation logic needed by clients, such as ASP. NET MVC in IIS web server.

c. Business tier: It handles and encapsulates all of business domains and logics; also called as domain layer.

d. Persistence tier: This tier handles the read/write of the business data to the data layer, also called data access layer (DAL).

e. Data tier: It is the external data source, such as a database.

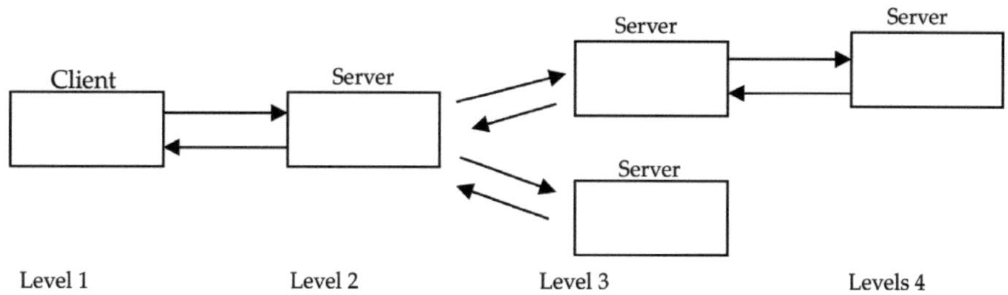

Figure 5. N-tier architecture.

4. Remote procedure mechanism

Remote procedure call works on client–server communication protocol that is used by one program to request a service from a program located in another computer in a network without understanding network details. It is based on RPC is a synchronous operation requiring the requesting program to be suspended till the results of remote procedure are returned [13].

4.1. Working and architecture of RPC

RPC is analogous to a function call extending the notion of conventional local procedure calling so that procedure need not exists in the same address space as the calling procedure. Like a function call, the calling arguments are passed to the remote procedure and the caller waits for a response to be returned from the remote procedure.

The client makes a procedure call that sends a request to the server and waits for response, as shown in **Figure 6**. The thread is blocked from processing until either a reply is received, or it times out. When the request arrives, the server calls a dispatch routine that performs the requested service, and sends the reply back to the client. After the RPC call is completed, the client program continues its normal execution [4].

Stub: Stubs are generated at the static compilation time and then deployed to the client side which is used as a proxy for the client. Client-side proxy acts as a mediator between the client and the broker and provides additional transparency between them and the client so that a remote object appears like a local one. The proxy hides the inter-process communication (IPC) at protocol level and performs marshaling of parameter values and un-marshaling of results from the server.

Skeleton: Skeleton is generated by the service interface compilation and then deployed to the server side, which is used as a proxy for the server. Server-side proxy encapsulates low-level

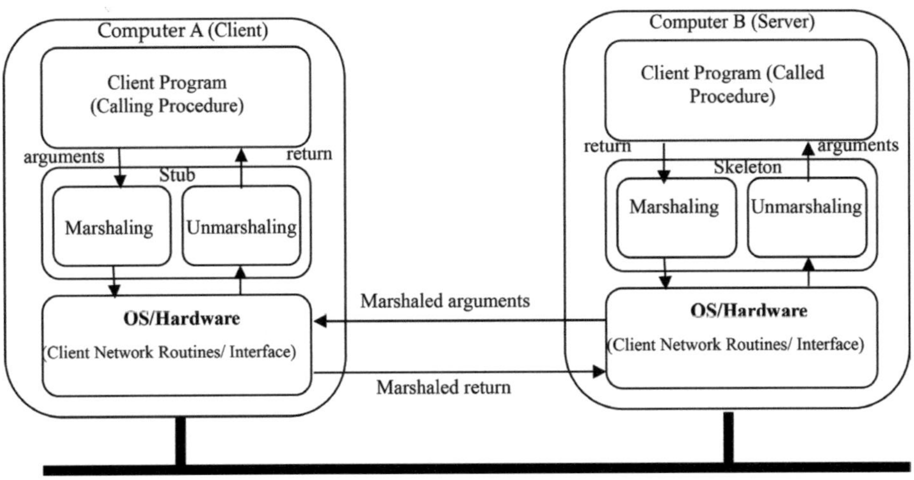

Figure 6. Remote procedure call.

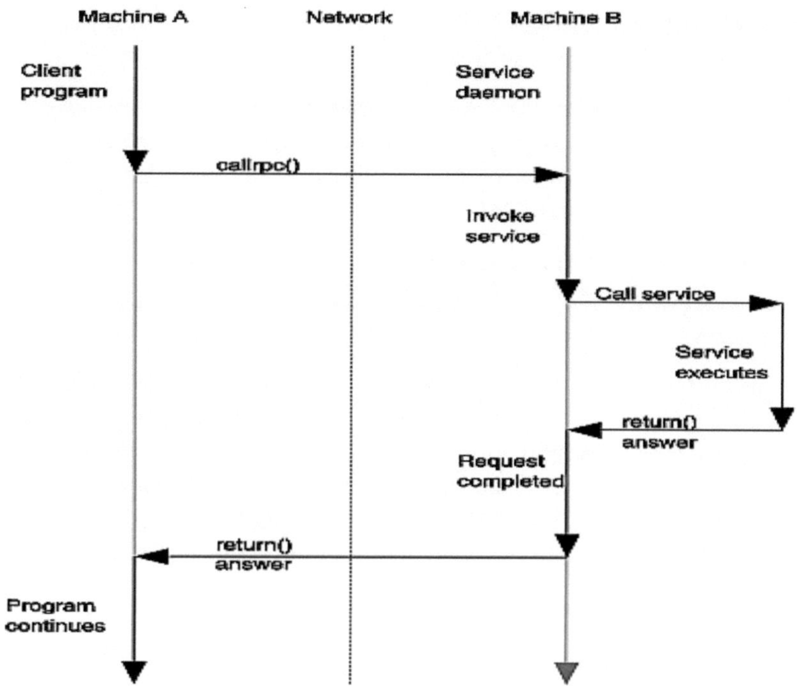

Figure 7. Event flow in RPC.

system-specific networking functions and provides high-level APIs to mediate between the server and the broker. It also receives the requests, unpacks the requests, un-marshals the method arguments, calls the suitable service, and also marshals the result before sending it back to the client [2].

Sequence of events during an RPC:

- The client calls the client stub. The call is a local procedure call, with parameters pushed on to the stack in the normal way.
- The client stub packs the parameters into a message and makes a system call to send the message. Packing the parameters is called marshaling.
- The client's local operating system sends the message from the client machine to the server machine.
- The local operating system on the server machine passes the incoming packets to the server stub.
- The server stub unpacks the parameters from the message. Unpacking the parameters is called un-marshaling.

Finally, the server stub calls the server procedure. The reply traces the same steps in the reverse direction. **Figure 7** shows the event flow of RPC.

5. Remote method invocation

Remote method invocation is a technology introduced by java that allows invocation of methods that are remotely located by simply calling them using desired interfaces. RMI technology allows us to distribute over business logic i.e. making the business logic available on a remote server letting it accessible to clients [14].

RMI is often called as "RPC with object orientation", i.e. the RPC but with ability to pass one or more objects along with the request. The objects can include the information that will change the service that is performed in the remote computer as delineated in **Figure 8**.

For example, when a user at a remote computer fills out an expense account, the Java program interacting with the user could communicate, using RMI, with a Java program in another computer that always had the latest policy about expense reporting. In reply, that program would send back an object and associated method information that would enable the remote computer program to screen the user's expense account data in a way that was consistent with the latest policy [15]. The user and the company both would save time by catching mistakes

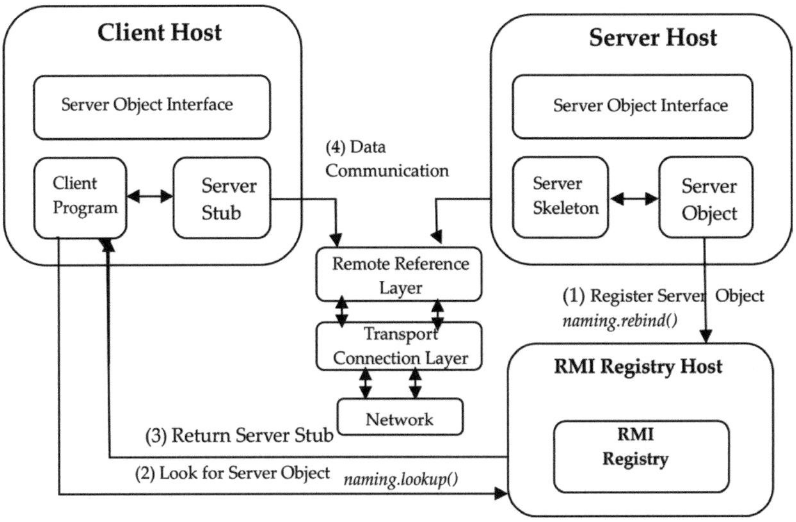

Figure 8. Remote method invocation.

RMI	RPC
Location neutral, language dependent	Language neutral mechanism
Supports object oriented design	It is procedural like C
It allows objects passing as arguments and return values	It supports only primitive data types
This allows usage of design patterns	No such capability

Table 1. RMI v/s RPC difference table.

http://dx.doi.org/10.5772/intechopen.80644

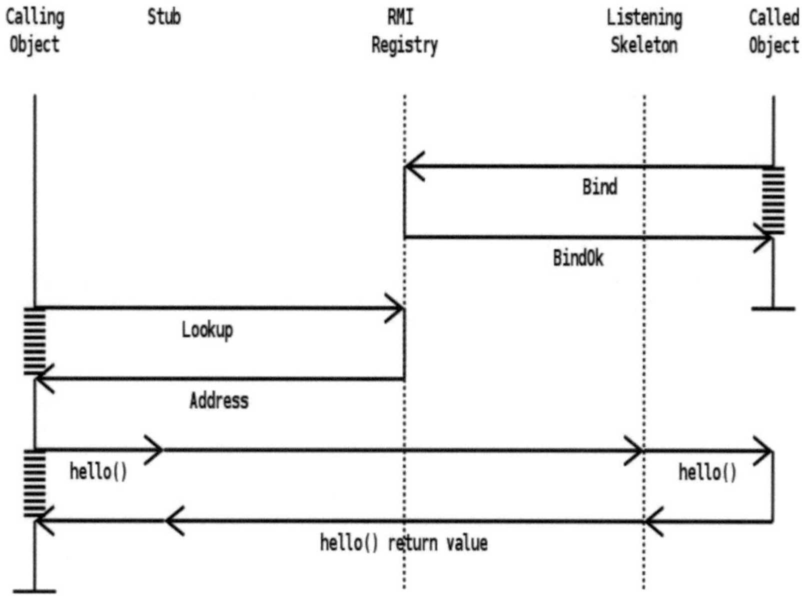

Figure 9. Event flow in RMI.

early. Whenever the company policy changed, it would require a change to a program in only one computer (**Table 1**).

RMI is implemented as three layers (as illustrated in **Figure 9**):

a. Stub/Skeleton layer: A stub program represents the remote object and also acts as gateway to a corresponding skeleton at the server end. The stub appears to the calling program to be the program being called for a service.

b. Remote reference layer: This can behave differently depending on the parameters passed by the calling program. For example, this layer can determine whether the request is to call a single remote service or multiple remote programs as in a multicast.

c. Transport connection layer: This sets up and manages the request. A single request travels down through the layers on one computer and up through the layers at the other end.

RMI Registry is a central repository keeping a track of all services being exposed from the current network. Since all the clients' requests for services through the RMI Registry the location of the application or service is unknown to the clients hence making the application location neutral [16].

6. Code-on-demand paradigm

Typically, code on demand is used for any technology that sends executable code from a server host to a client host on the request of the client's application. Code on demand is a specific use of mobile code under the field of code mobility. In the code-on-demand style, as delineated in **Figure 10**, a client component has an access to a set of resources, but not the know-how on how to process them. It sends a request to a remote server for the code representing that know-how, receives that code, and executes it locally. So as per the code-on-demand paradigm, knowing the know-how is necessary when in need [17].

Say for example, one host (A) initially is unable to execute its task due to a lack of code (know-how). And another host (B) in the network provides the needed code. Once the code is received by A, the computation is carried out on A's machine. Host A holds the processor capability as well as the local resources. Unlike in the client–server paradigm, A does not need knowledge about the remote host, since all the necessary code will be downloaded.

Java applets are excellent practical examples of this paradigm. Applets get downloaded in Web browsers and execute locally.

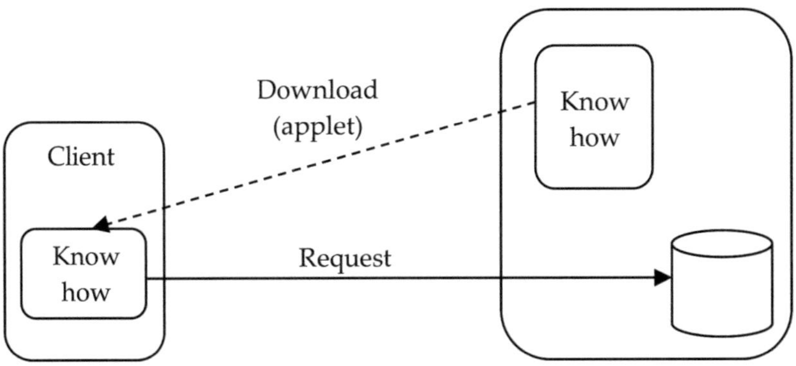

Figure 10. Code-on-demand.

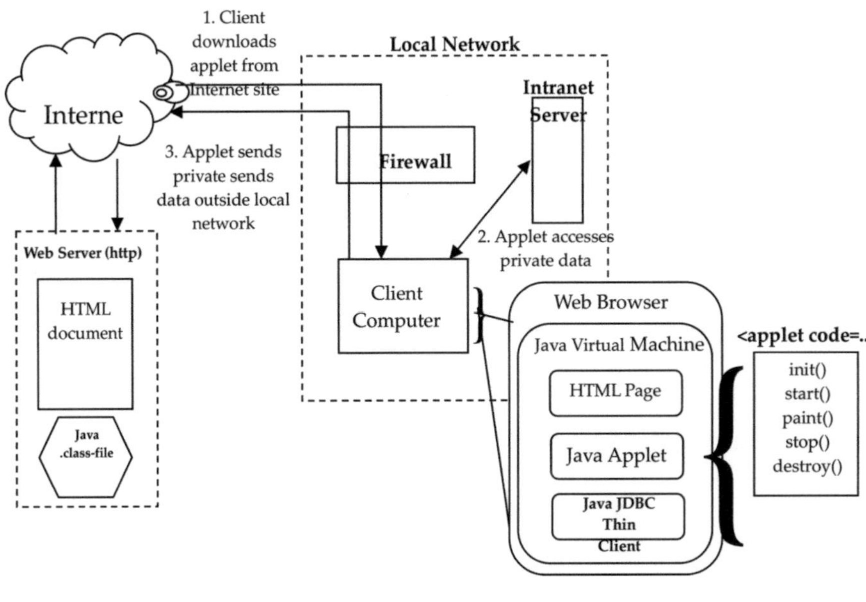

Figure 11. Architecture of applet.

http://dx.doi.org/10.5772/intechopen.80644

6.1. Working and architecture of applets

The internet is a combination of various kinds of systems or platforms that are often required to communicate with each other. The client that makes a request may be from a completely different platform for instance the application may be hosted on the windows based server and client may be requesting from a Linux-based system.

Java introduced a new technology that would allow any client from any network platform to host and execute applications over the internet. This new technology was called as applets [18]. The word applet stands for an "application scriplets". This can be defined as a piece of java code residing on a server machine requested via a browser downloaded over the internet and executed on the client machine via the browser. In order to execute the applet on a client machine, the browser must be java enabled i.e. JRE must be enabled. An applet is typically embedded inside a web page and runs in the context of a browser. The browser's Java Plug-in software manages the lifecycle of an applet. The architecture of applet is shown in above **Figure 11**.

6.2. Life cycle of an applet

Atop these five methods, depicted in **Figure 12**, an applet is been created:

a. Init(): This method is intended for whatever initialization is needed for your applet. It is called after the param tags inside the applet tag have been processed.

b. Start(): This method is automatically called after the browser calls the init method. It is also called whenever the user returns to the page containing the applet after having gone off to other pages.

c. Stop(): This method is automatically called when the user moves off the page on which the applet sits. It can, therefore, be called repeatedly in the same applet.

d. Destroy(): This method is only called when the browser shuts down normally. Because applets are meant to live on an HTML page, you should not normally leave resources behind after a user leaves the page that contains the applet.

e. Paint(): Invoked immediately after the start() method, and also any time the applet needs to repaint itself in the browser. The paint() method is actually inherited from the java.awt.

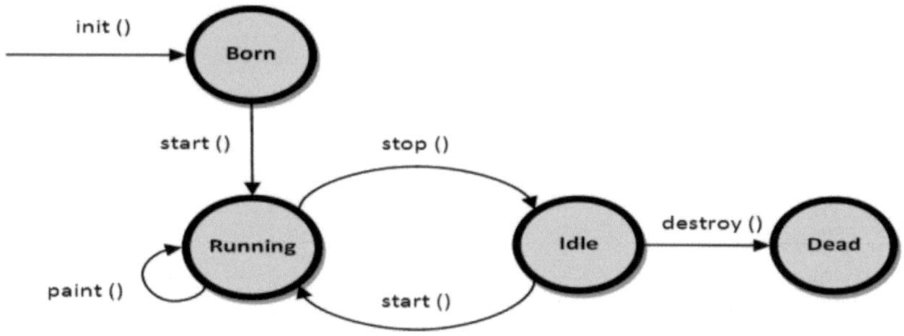

Figure 12. Life cycle of applet.

7. Remote evaluation

In computer science, remote evaluation is a term that belongs to the family of mobile code, within the field of code mobility. It is for any technology that involves the transmission of executable software code from a client hosts to a server hosts for execution to be happen at the server and the result is sent back to the client after execution for this resources of server sside are used [19]. A simple model of remote evaluation is illustrated in **Figure 13**.

An example for remote evaluation is grid computing: An executable task may be sent to a specific computer in the grid. After the execution has terminated, the result is sent back to the client. The client in turn may have to reassemble the different results of multiple concurrently calculated subtasks into one single result.

7.1. Working and architecture of servlets

7.1.1. Basic idea and architecture

Web based technologies are of two different types: Client Side Technologies and Server Side Technologies. The Client Side Technology has the code completely downloaded on the client machine and executed on the client itself, any changes that need to be incorporated or updated in the application will be on client system after re-downloading by the client. The processing of this application will take place on the client, completely.

In a Server Side Technology the complete business logic is maintained on the server and on the request of the client it will be executed on the server, delivering the response to the clients. The Java Servlets technology provides on such simple, vendor-independent mechanism to extend the functionality of a web server [20]. Servlets technology is similar to common gateway interface (CGI) scripts, Javascripts (on client side) and hypertext preprocessor (PHP). Additionally, scripting languages can be used in servlets to dynamically modify or generate hypertext markup language (HTML) pages. It also supports various HTTP methods, such as GET and POST, which is used to redirect requests and responses as shown in **Figure 14**.

7.1.2. Working and life cycle

Whenever a client sends a request to the J2EE application server for a particular servlet, the J2EE Application server passes the request to the Web container. The Web container checks

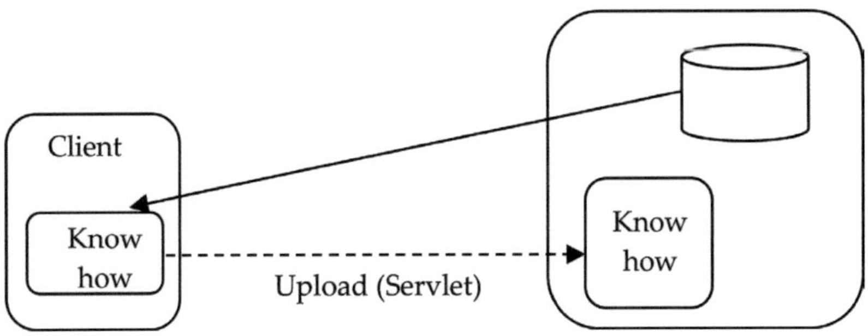

Figure 13. Remote evaluation.

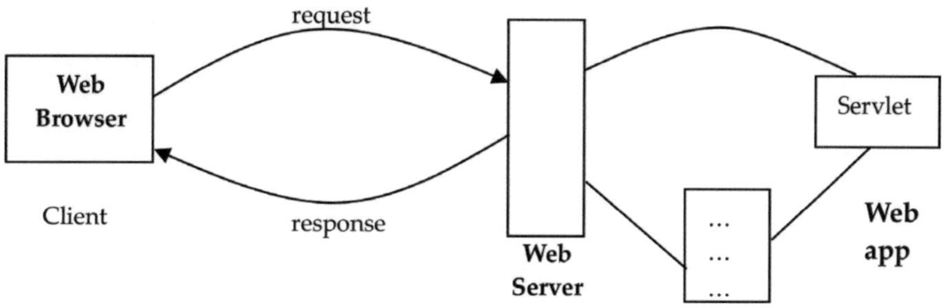

Figure 14. Architecture of servlets.

whether an instance of the requested servlet exists. If the servlet instance exists then the Web container delegates the request to the servlet, which process the client request and sends back the response (Shown in **Figure 15**).

It is the job of Web container to get the request and response to the servlet. The container creates multiple threads to process multiple requests to a single servlet. So in case the servlet instance does not exist, the Web container locates and loads the servlet class. The Web container then creates an instance of the servlet and initializes it. The servlet instance starts processing the request after initialization. The Web container passes the response generated by the servlet to the client.

Servlets don't have a main() method that's why Web container manages the life cycle of a Servlet instance. The life cycle of the servlet includes three states: new, ready and end. The servlet is in new state if servlet instance is created. After invoking the init() method, Servlet comes in the ready state [21]. In the ready state, servlet performs all the tasks. When the web container invokes the destroy() method, it shifts to the end state. It is shown in **Figure 16**.

7.1.3. Life cycle of servlet

a. Servlet class is loaded: The class loader is responsible to load the servlet class. The servlet class is loaded when the first request for the servlet is received by the web container.

b. Servlet instance is created: The web container creates the instance of a servlet after loading the servlet class. The servlet instance is created only once in the servlet life cycle.

c. Init method is invoke: The web container calls the init method only once after creating the servlet instance. The init method is used to initialize the servlet.

   Method Signature: public void init(ServletConfig config) throws ServletException

d. Service method is invoked: The web container calls the service method each time when request for the servlet is received [22]. If servlet is not initialized, it follows the above three steps then calls the service method. The servlet is initialized only once so if servlet is already initialized, it directly calls the service method.

   Method Signature: public void service(ServletRequest request, ServletResponse response) throws ServletException, IOException.

e. Destroy method is invoked: The web container calls the destroy method before removing the servlet instance from the service. It gives the servlet an opportunity to clean up any resources like memory, thread etc. **Figure 16** shows life cycle methods of servlets.

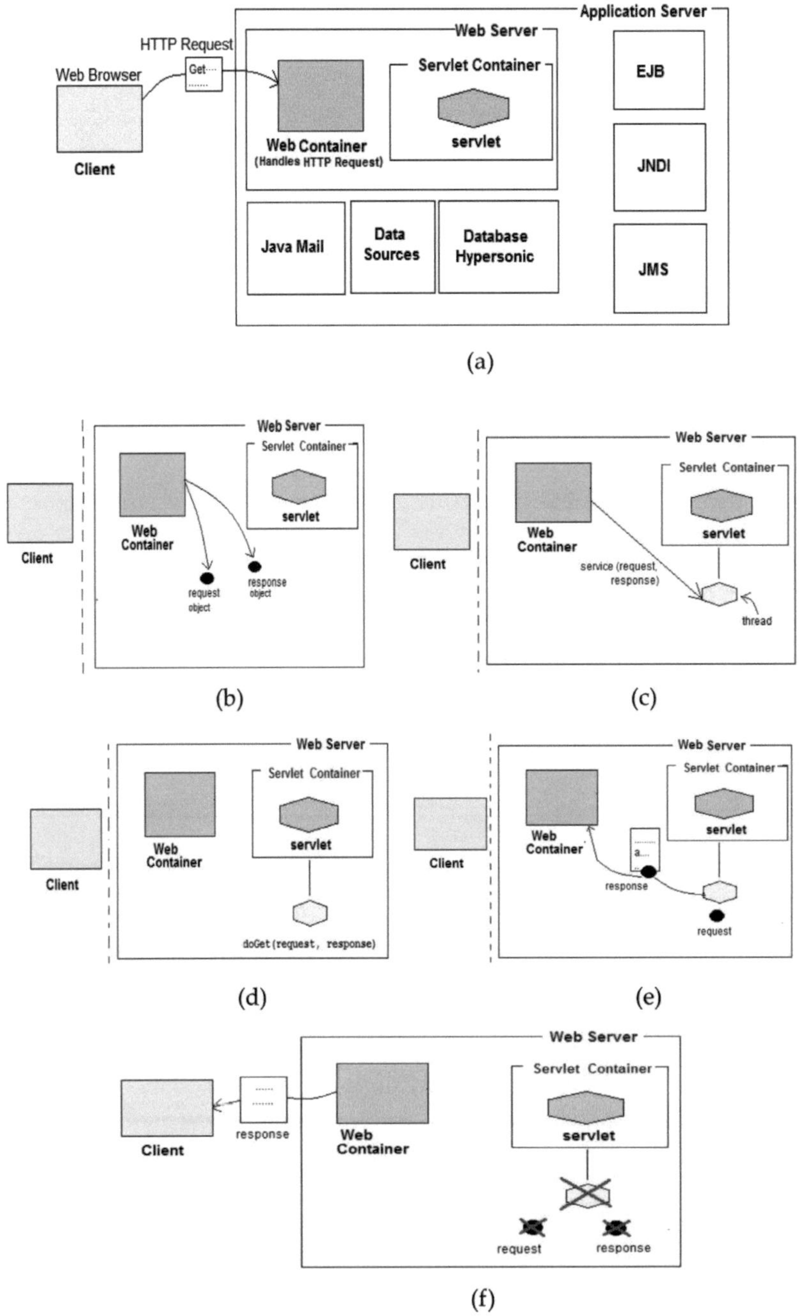

Figure 15. Various stages in request and response mechanism of servlets. (a) Clients request handling carried out by web container. (b) Object formation. (c) Calling servlet thread. (d) Thread execution. (e) Submission of response. (f) Final response toclient.

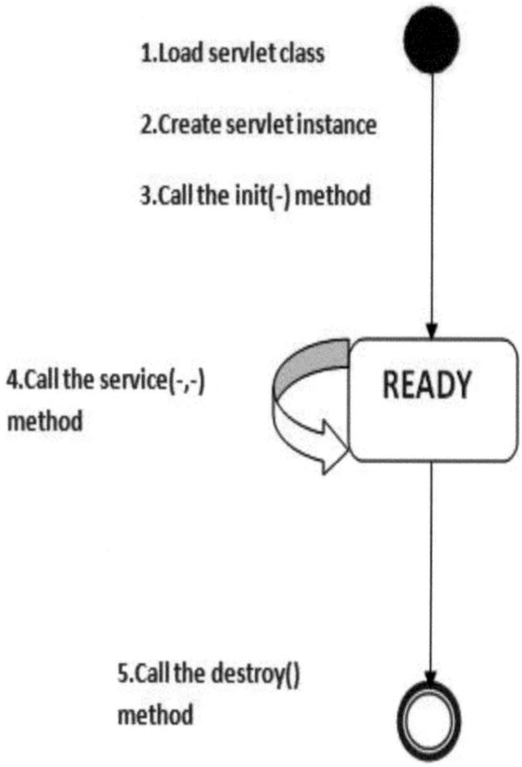

Figure 16. Life cycle of servlets.

8. Conclusion

In the past couple of years there has been a development of enthusiasm for versatile platform innovation and a few stages have been developed and innovated to allow more independencies in programming platform. In this chapter we have surveyed and researched the various computing environment provided for remote execution that has incurred the need of mobile codes, intelligent agents, autonomous objects, etc. raising issues with flexibility, efficiency and security in present system that can promises to resolve existing problems and add on more facilities like remote execution, auto-scheduling and many more. Some of them have just been utilized for look into purposes while others have been conveyed as business items. A few technologies that incorporated in evolution of mobile agents have been discussed on the basis of the usefulness of have been displayed in this exploration.

Acknowledgements

I would like to impart my sincere thanks to my guide Late. Dr. Rajesh Purohit, who has inspirited and fostered my interest in multifarious streams and disciplines of remoting and mobile-objects. Further I would like to extend my regards to all academic friends and lecturers who supported and motivated to move on with my work. They are (alphabetical order) Ashish Sharma, Poonam Purohit, Purva Dayya and Shivam Lohiya.

The following presents the acronyms used throughout this chapter.

Abbreviations

HSD	hosted shared desktop
VDI	desktop virtualization
CORBA	common object request broker architecture
WBEM	web based enterprise management
MbD	management-by-delegation
REV	remote revaluation
CoD	code on demand
TCP/IP	transfer control protocol/internet protocol
PL/UI	presentation layer
BLL	business logic layer
DAL	data access layer
API	application programmable interface
UI	user interface
WPF	windows presentation foundation
HTML	hyper text markup language
RPC	remote procedure call
RMI	remote method invocation
Applets	application scriplets
Servlets	server scriplets
CGI	common gateway interface
HTTP	hyper text transfer protocol
PHP	hypertext pre processor

Author details

Rahul Singh Chowhan

Address all correspondence to: word2rahul@gmail.com

Agriculture University, Jodhpur, India

References

[1] Aridor Y, Lange DB. Agent design patterns: Elements of agent application design. In: Proceedings of the Second International Conference on Autonomous Agents; 1 May 1998. ACM. pp. 108-115

[2] Juziuk J. Design patterns for multi-agent systems. Linnaeus University, Faculty of Science and Engineering, School of Computer Science, Physics and Mathematics; 2012

[3] Pandey R, Sharma N, Rathore R. Aglets (A java based Mobile agent) and its security issue. International journal of emerging trends & technology in computer science (IJETTCS). 2013;**2**(4):107-114

[4] Ahila SS, Shunmuganathan KL. Overview of mobile agent security issues—Solutions. In: 2014 International Conference on Information Communication and Embedded Systems (ICICES); 27 Feb 2014. IEEE. pp. 1-6

[5] Bertsekas, Dimitri P. Dynamic programming and optimal control. Vol. 1. Belmont, Massachusetts: Athena Scientific; 1996

[6] Puterman ML. Markov decision processes: Discrete stochastic dynamic programming. Wiley Series in Probabilty and Statistics. Hoboken, New Jersey: John Wiley & Sons, Inc.; 2014

[7] Bellman RE, Dreyfus SE. Applied dynamic programming. Princeton Legacy Library. London: Oxford University Press, Princeton University Press; 2015 Dec 8:**2050**

[8] Wang D, Mu CX, Liu DR. Data-driven nonlinear near-optimal regulation based on iterative neural dynamic programming. Zidonghua Xuebao/Acta Automatica Sinica. 2017 Mar;**43**(3):366-375

[9] Ahmed K, Bigagli D, Hu Z, Wang J. Inventors; International Business Machines Corp, assignee. Resource Manager for Managing the Sharing of Resources Among Multiple Workloads in a Distributed Computing Environment. United States Patent US 9,632,827; 2017 Apr 25

[10] Kurniawan B. Java for the Web with Servlets, JSP, and EJB. New Riders Publishing; 2002

[11] Siek K, Wojciechowski PT. Atomic RMI: A distributed transactional memory framework. International Journal of Parallel Programming. 2016 Jun;**44**(3):598-619

[12] Czaja L. Remote procedure call. In: Introduction to Distributed Computer Systems. Cham: Springer; 2018. pp. 141-155

[13] Kaur M, Sharma S. A dynamic clone approach for mobile agent to survive server failure. In: 2015 4th International Conference on Reliability, Infocom Technologies and Optimization (ICRITO) (Trends and Future Directions); 2 Sep 2015. IEEE. pp. 1-5

[14] Diaz J, Munoz-Caro C, Nino A. A survey of parallel programming models and tools in the multi and many-core era. IEEE Transactions on Parallel and Distributed Systems. 2012 Aug;**23**(8):1369-1386

[15] Hwu WM, Ryoo S, Ueng SZ, Kelm JH, Gelado I, Stone SS, Kidd RE, Baghsorkhi SS, Mahesri AA, Tsao SC, Navarro N. Implicitly parallel programming models for thousand-core microprocessors. In: Proceedings of the 44th annual Design Automation Conference; 4 Jun 2007. ACM. pp. 754-759

[16] González-Vélez H, Leyton M. A survey of algorithmic skeleton frameworks: High-level structured parallel programming enablers. Software: Practice and Experience. 2010 Nov;**40**(12):1135-1160

[17] Manogaran G, Lopez D. Health data analytics using scalable logistic regression with stochastic gradient descent. International Journal of Advanced Intelligence Paradigms. 2018;**10**(1-2):118-132

[18] Haefner JW. Parallel computers and individual-based models: An overview. In: Individual-based models and approaches in ecology. Chapman and Hall/CRC; 2018. pp. 126-164

[19] Murphy R, Sterling T, Dekate C. Advanced architectures and execution models to support green computing. Computing in Science & Engineering. 2010 Nov;**12**(6):38-47

[20] Marowka A. On parallel software engineering education using python. Education and Information Technologies. 2018 Jan;**23**(1):357-372

[21] Kemp R. Current developments in open source software. Computer Law and Security Review. 2009 Nov 1;**25**(6):569-582

[22] Coulthard P, Yantzi DJ, Simpson EV. Inventors; International Business Machines Corp, assignee. Framework to Access a Remote System from an Integrated Development Environment. United States patent US 8,296,720; 2012 Oct 23

Chapter 3

A Nondestructive Distributed Sensor System for Imaging in Industrial Tomography

Tomasz Rymarczyk

Additional information is available at the end of the chapter

http://dx.doi.org/10.5772/intechopen.79567

Abstract

The proposed solution is based on the construction of a cyber-physical system for acquiring, processing, and reconstructing the image of measurement data. Industrial tomography enables observation of physical and chemical phenomena without the need of internal penetration. Process tomography gives the possibility to analyze processes taking place inside the facility without disturbing the production, analysis, and detection of obstacles, defects, and various anomalies. The presented measuring system has a specially designed measuring structure (including electrodes, thanks to which it is an innovative solution in the field, particularly effective in analysis). Knowledge of the characteristics of each tomographic technique allows to choose the appropriate method of image reconstruction. The inverse problem is the process of identifying optimization or synthesis, wherein the objective is to determine the parameters describing the data field.

Keywords: process tomography, electrical impedance tomography, inverse problem, electrical capacitance tomography, ultrasound tomography

1. Introduction

The system is based on a tomography in a cyber-physical system (CPS) of self-control. Active control-optimized functions can only be implemented using a system that allows electronic control. The algorithms concern issues related to the processing of data obtained from various sensors located in nodes. Monitoring takes place within the scope of acquired and processed data and the automation of parameters. Advanced automation and control of production processes play a key role in maintaining the competitiveness of the economy. While costly process equipment and production lines can be seen as the heart of industrial production, control systems and information technology are his brain. They provide flexibility to quickly adapt production processes to changing customer requirements and ensure safety and performance at the lowest possible cost of resources and energy. Therefore, developing and applying advanced process control is one of the most effective levers for immediate and long-term energy savings, better product quality, increased process security, and higher production flexibility and will ensure and stimulate business development in conventional and emerging areas, create new jobs, and improve social standards in Europe. In addition, cyber-physical systems (CPS) are physical and engineering systems whose activities are monitored, coordinated, controlled, and integrated by the computing and communication core. CPS combines discrete and powerful computational logic to monitor and control the continuous dynamics of physical and engineering systems. Precision of calculations must be connected with uncertainty and noise in the physical environment. The lack of perfect synchronization in time and space must be solved. Component errors in cyber and physical domains must be tolerated or restricted. These needs require the creation of innovative science and engineering principles. The trial-and-error approach to building computer-oriented systems must be replaced by rigorous methods, certified systems, and advanced tools. Security and privacy requirements must be enforced. The system dynamics should be addressed on many time scales. The scale and growing complexity must be tame [1, 2].

Tomography is a technique of imaging the interior of a tested object using measurements made on the edge. Depending on the technological specifics, you can see both advantages and disadvantages in terms of accuracy, frequency, and resolution of reproduced images. In order to get information about the object being studied, it uses a variety of physical phenomena that are carriers of information: X-rays, gamma rays, ultrasounds, electron beams, electric currents, magnetic fields, and photons [3, 4]. Knowledge of the characteristics of each tomographic technique allows to choose the appropriate method of image reconstruction [5–7].

The main purpose of this work is to design a system for data acquisition and analysis by reconstructing the image for various tomographic methods [8]. Control methods include issues related to the processing of data obtained from various sensors located in nodes. Monitoring takes place as part of the data processing and parameters obtained and processed. Multiphase flow measurement technologies are still being built and improved. There is a clear tendency in the industry to implement more optimal related functions with an emphasis on active inspection and monitoring.

Although there are many methods to optimize technological processes, there is no universal solution that would be optimal in a wide range of measurement conditions. A new hybrid solution was proposed using imaging techniques together with appropriate sensors. This article describes several types of algorithms and models of reconstruction. The solution to this optimization problem is obtained by combining several numerical methods. The reconstruction of 2D examples with the use of numerical and experimental data is shown. The proposed tomographic system consists of a set of sensors (devices) and software that uses Cloud Computing and the Big Data cluster to process, visualize, and analyze data (cyber-physical system) [9].

2. System architecture

The article presents a model of intelligent platform enabling configuration and cooperation with various external systems. The platform enables the management of a smart company structure in terms of processes, products, simulations, and virtual products. This allows optimization and automatic optimization of production processes. It will also allow you to follow the production cycle and ensure cooperation with external applications. The system can operate autonomously, monitor, control, collect, and collect measurement data. The open platform model of intelligent devices and sensor technology based on tomography in the cyber-physical self-monitoring system includes new measurement techniques and the construction of innovative intelligent measuring devices, system structure along with communication interface, unique algorithms for data optimization and analysis, image reconstruction algorithms, and monitoring technological processes.

The system consists of the following components (see **Figure 1**):

- New measurement techniques and designing of innovative devices and measuring sensors
- Algorithms of processes of reconstruction and monitoring of processes
- New unique algorithms for data optimization and analysis
- System structure and communication interface
- Cyber-physical platform.

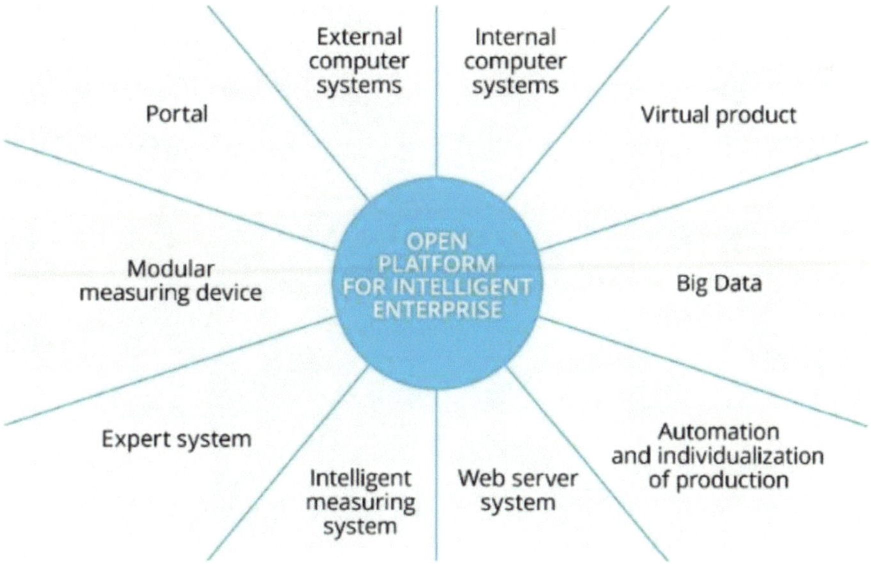

Figure 1. The idea of the system.

The system allows management of the intelligent enterprise structure in terms of processes and simulations. **Figure 2** shows the model of the system for optimization and automation of production processes with data analysis and product virtualization. Individualization of the system in the cooperative model is presented in **Figure 3**. It will also allow to follow the production cycle and ensure cooperation with external applications.

Wireless technology has been used as the first open wireless standard to control industrial processes in sensor networks. In order to monitor the infrastructure of linear and hierarchical projects, various wireless transmission standards can be applied at different levels of the hierarchy. In the tomography system, we often use the MQTT (Message Queue Telemetry Transport) protocol. Many standards of wireless networks, such as Bluetooth, ZigBee, and DASH7, are often used in integration to implement sensor networks. In the case of pipelines located in remote areas, it is desirable to use long-distance networks such as GSM and GPRS for sending data collected in the backhaul network. After successful measurement and registration of parameters from the sensors, a mechanism for sending data to the base station is required. Reliable and secure data transmission is extremely important. For this purpose, various network architectures and topologies can be used. Factors such as designing real-time detection nodes, pipeline and network infrastructure, connecting nodes with the base station, and battery life or work cycle directly affect data transmission.

The algorithms based on tomographic techniques can be used to automatically control and solve problems related to the processing of data obtained from various sensors located in key nodes of the system. Supervision and control are in the scope of acquired and processed data and parameters of devices implementing automation, such as servo valves, pump supply,

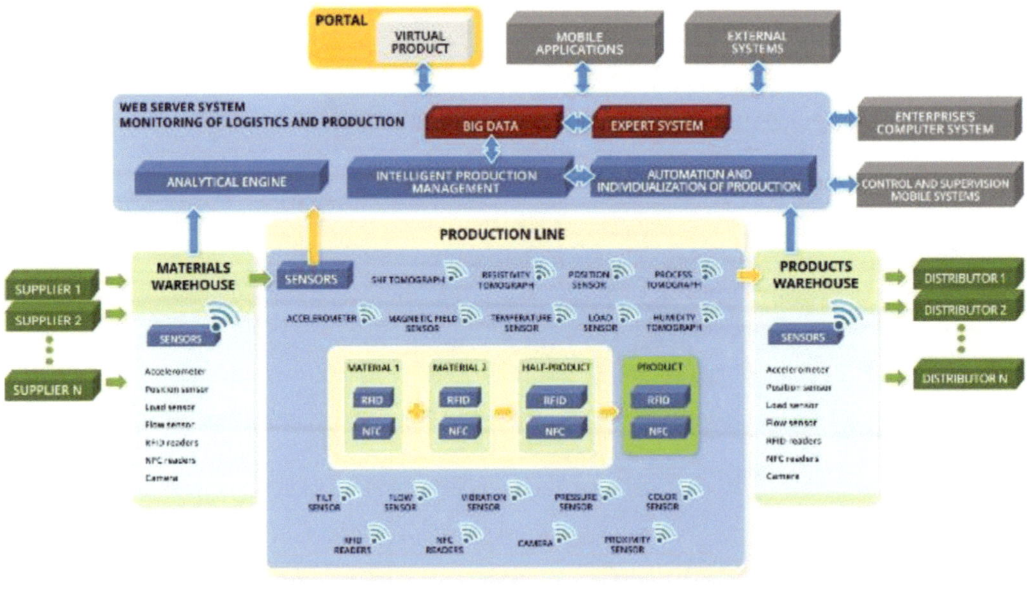

Figure 2. Model of functional system.

http://dx.doi.org/10.5772/intechopen.79567

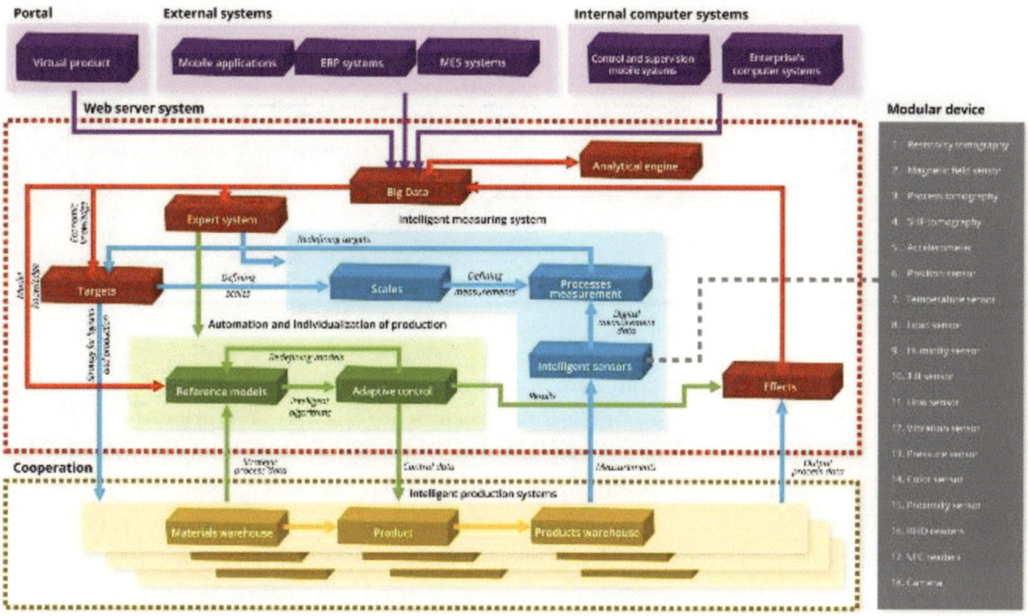

Figure 3. Model of production processes.

and rotary flow. The basic feature of using wireless methods is to obtain important information about the process and the state of installation in real time [10–13].

3. Process tomography

Process tomography gives the opportunity to analyze the processes taking place inside the facility without interfering with the production, analysis, and detection of obstacles, defects, and various anomalies. The tomography belongs to the opposite problems of the electromagnetic field. The inverse problem is the process of identification, optimization, or synthesis, in which the goal is to determine the parameters describing the data [14–16]. Process tomography aims to determine the properties of the tested object from measurements at its edge [17, 18].

Tomography of industrial processes is a harmless, noninvasive imaging technique used in various industrial technologies in processes. It plays an important role in continuous measurement data, which allows better understanding and monitoring of industrial processes, providing a fast and dynamic response, which facilitates process control in real time detecting failures and abnormal system operation. Tomographs allow us to "look inside" pipes in flow reactors. Industrial tomography enables observation of physical and chemical phenomena without the need of internal penetration. In order to obtain information about the test object, measurements are used in various physical phenomena, in which information carriers are X-rays, gamma rays, ultrasounds, electron beams, electric currents, and magnetic fields. The main advantage of tomographic examinations is their noninvasiveness in the

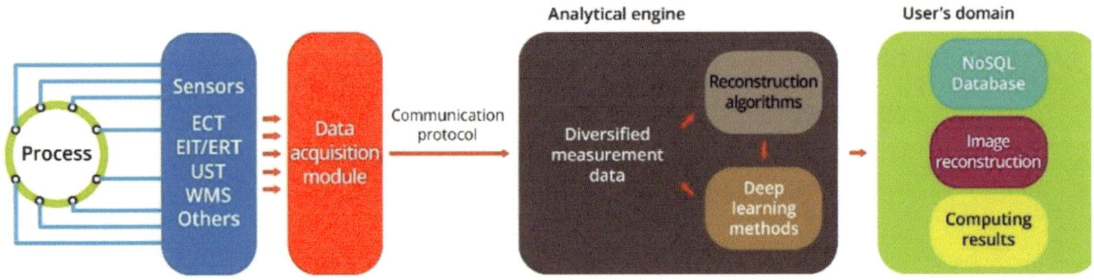

Figure 4. General measurement model for tomography sensors.

studied environment. Such measurements do not change physical and chemical parameters. In the reconstruction of the image, the key parameters are the speed of analysis of flowing raw materials and the accuracy of reconstructed processes. The measurement must be fast because some industrial processes run at high speed. The measuring system consists of a sensor, specialized electronics for capacitance measurement, and data reconstruction and analysis system. Industrial tomography applications are usually a challenge for obtaining spatial distribution data from observations that go beyond the process boundary. The biggest challenge is to achieve effective coverage of closed spaces using practical resources at a reasonable cost. Sensor networks with feedback loops are the basic elements of production control. Distributed infrastructure requires various tasks related to detection and startup and is usually characterized by internal spatial organization. The decisive difference in the mass production of chemicals, food, and other goods is that joint process sensors only provide local measurements. In most production systems, such local measurements are not representative of the entire process; therefore, spatial solutions are necessary. The general measurement model for tomographic sensors is shown in **Figure 4**.

4. Methods and models

4.1. Electrical tomography

The solution to the electrical impedance tomography (EIT) problem is to determine the potential distribution in the region Ω in given boundary conditions and full information about the analyzed region [19–21]. In order to obtain benefits, the accuracy of EIT has historically been divided into capacity (ECT), for systems dominated by dielectrics and resistance (ERT), for conducting processes. The basic theory can be obtained from Maxwell's equations. A complex "admittivity" is defined as follows:

$$\gamma = \sigma + i\omega\varepsilon \tag{1}$$

where σ is the electrical conductivity, ε is the permittivity, and ω is the angular frequency. For an electric field strength (E), the free current density (J) in the area under investigation will be related by Ohm's law:

$$J = \gamma E \tag{2}$$

The gradient of the potential distribution (u) is as follows:

$$E = -\nabla u \tag{3}$$

Assuming there are no current sources within the examined region then from Ampère's law:

$$\nabla \cdot J = 0 \tag{4}$$

The potential distribution in the isotropic, inhomogeneous region is as follows:

$$\nabla \cdot (\gamma \ \nabla u) = 0, \tag{5}$$

where u is the potential.

As above the ratio of $\omega\varepsilon/\sigma$, when the capacitance or the resistance term is dominant, the governing equation is further simplified:

$$\nabla \cdot (\sigma \ \nabla u) = 0 \, for \, \frac{\omega\varepsilon}{\sigma} \ll 1 \, (\text{ERT}) \tag{6}$$

$$\nabla \cdot (\varepsilon \ \nabla u) = 0 \, for \, \frac{\omega\varepsilon}{\sigma} \gg 1 \, (\text{ECT}) \tag{7}$$

As a result of the inverse problem solution, the conductivity distribution in the tested area is obtained.

A set of electric currents are injected into the examined object through these electrodes, and the obtained voltages are measured using the same electrodes. **Figure 5** shows the opposite method of acquiring boundary potential data illustrated for a cylindrical volumetric guide and 16 equally spaced electrodes: (a) first measurement and (b) second measurement.

In electrical capacitive tomography, the information source is the electrical capacitance between the electrodes located on the perimeter of the measurement sensor (see **Figure 6**). An important feature of the measurement is the non-invasive contact of the sensor with the tested object. Such a solution does not interfere with industrial processes. The advantage of this technique is the quick acquisition of measurement data. The laboratory measuring system with sensors is shown in **Figure 7**.

4.2. Ultrasound tomography

Measurement methods using information contained in the ultrasonic signal after passing through the medium under investigation are called ultrasound transmission methods (see **Figure 8**). Ultrasonic waves belong to short waves and have propagating and radiation properties. The length of these waves depends on the medium to which they are emitted

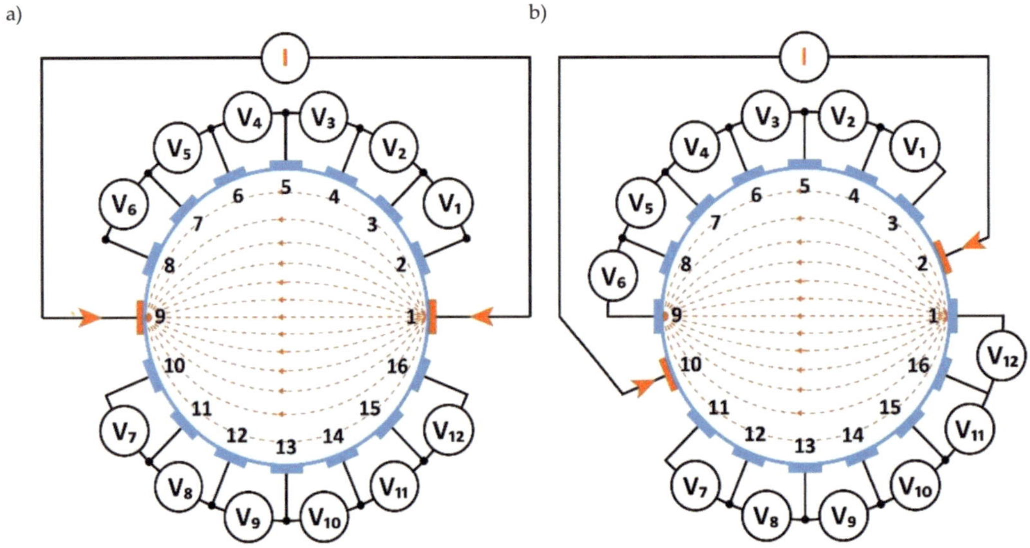

Figure 5. Opposite method in electrical resistance tomography.

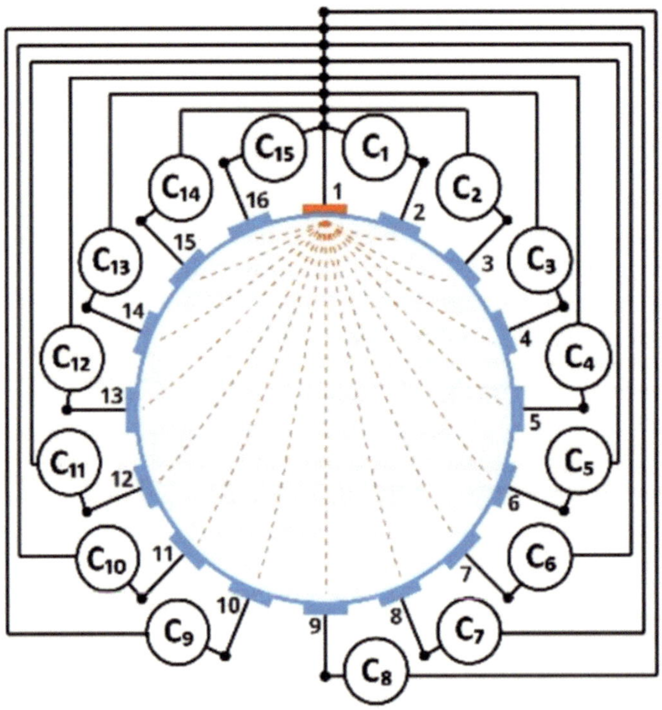

Figure 6. Measurement model of electrical capacitance tomography.

Figure 7. Laboratory measurement system.

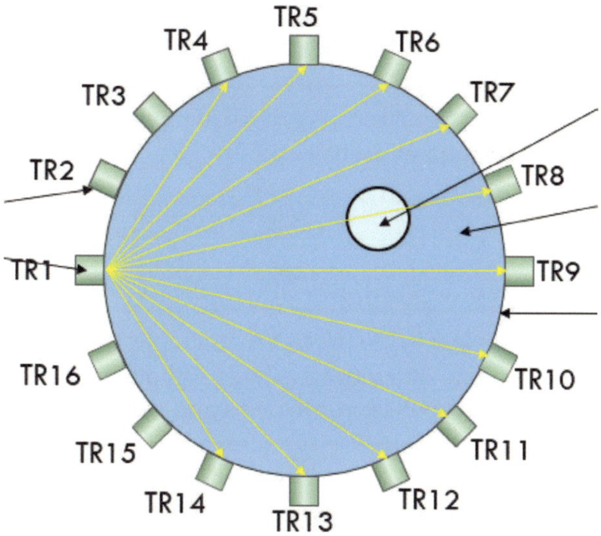

Figure 8. Idea of measurement model in ultrasound tomography.

and is in the range from a few micrometers in liquids to several dozen centimeters in metals. They can be used to measure the attenuation coefficient and ultrasonic time of signal transitions in the medium subjected to their influence. Many measurements can be made using ultrasound without fear of damage or irradiation of the examined objects. This technique allows to obtain quantitative images of the internal structure, in which the numerical values of each pixel describe such physical properties of the examined objects, e.g., temperature distribution, density, or viscosity. Measurements of parameters such as signal transitions, attenuation coefficient, and its derivative through frequency allow after appropriate reconstructive transformations, imaging of the internal structure of the tested medium, as well as flow parameters such as its speed, average speed, or profile speed. The basis for imaging is differences in local values of specific acoustic parameters. The image obtained by means of appropriate reconstruction methods presents the distribution (obtained from the measurement data using the scanning technique from as many directions as possible after the ultrasonic pulses have passed through the tested environment) [22].

The problem of image construction in the case of ultrasound very often leads to the overdetermined algebraic set of equations that can be expressed in the matrix form:

$$Wf = s, \tag{8}$$

where W is the matrix of dimensions m x n and m > n, s is the right-hand side vector (one column matrix), and f is the solution vector.

One of the ways of the solution (Eq. (7)) is to find the vector f^*, which minimizes Euclidean norm of residual vector r for the known matrix W and vector s, and it means:

$$\|r\|_2 = \min \|s - Wf\|_2, \quad \|f^*\|_2 = \min \|f\|_2 \tag{9}$$

where the last minimum is taken for all vectors f which fulfill the previous relation. Equation (8) is well known as a linear least squares problem (LSP).

5. Results

Numerical analysis of the problem in the process tomography takes place using, among others, finite element methods, finite difference methods, or boundary element methods. In the case of data shortages, we talk about fixed problems and in the case of excess with overdetermined problems. Automatic data analysis is an important part of the diagnosis of the process based on tomography. Knowledge of the process can make image reconstruction more resistant to incomplete or corrupted data. As a result of the calculations, we obtain a reconstructed image, for example, acoustic impedance, permittivity or conductivity, and dielectric loss to the parameters of a physical process (phase embolism, density, type of concentration). Advanced analysis leads to the extraction of features.

The example of image detection is shown in **Figure 9**. The images show the image reconstruction achieved by the Gauss-Newton method. A mesh of finite elements has been generated inside the area. The algorithm is solving for such a distribution of conductivity, so that the values of interelectrode voltages calculated on its basis are as close as possible to the measuring values of these voltages.

Figure 10 presents the image reconstruction by electrical capacitance tomography using the Levenberg-Marquardt method and the modified Levenberg-Marquardt method. Grids used in calculations: rare 2218 nodes and 4146 finite elements and dense 7861 nodes and 15,146

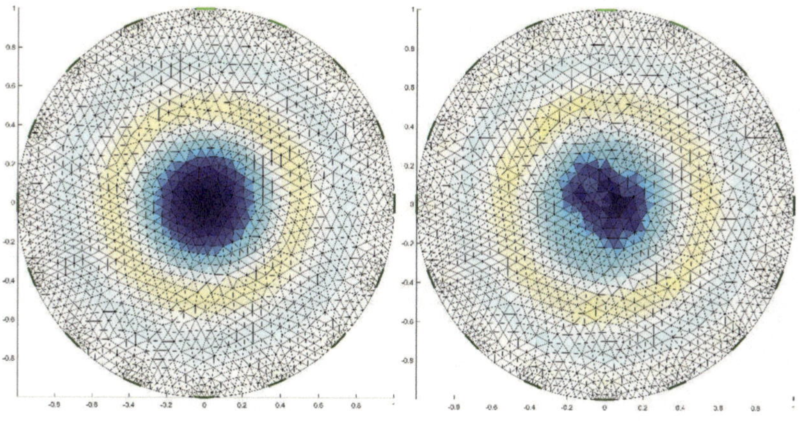

Figure 9. Image reconstruction (ERT): (a) Gauss-Newton with Laplace regularization and (b) Gauss-Newton with Tikhonov regularization (ERT).

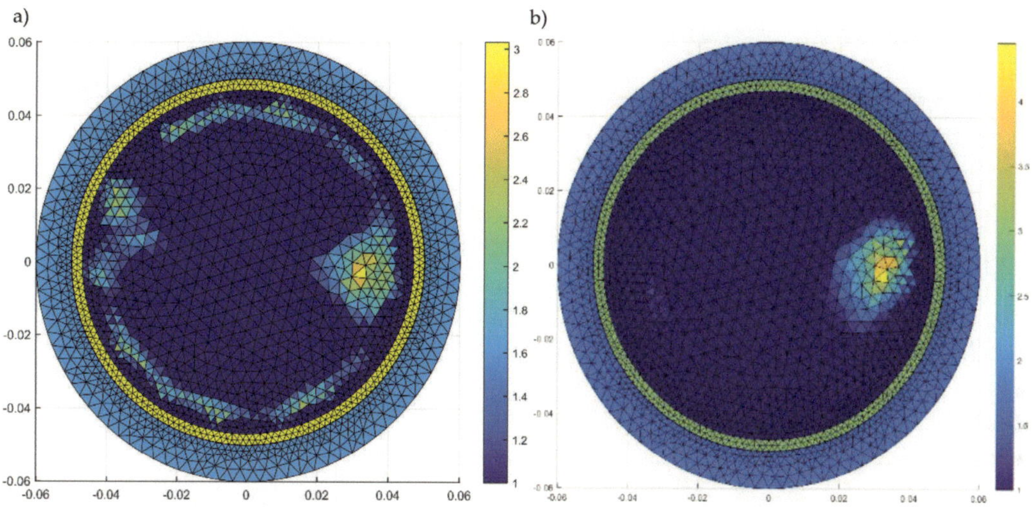

Figure 10. Example of the image reconstruction (ECT). (a) Levenberg-Marquardt method and (b) modified Levenberg-Marquardt method.

finite elements. The numerical experiment was carried out on noisy data. **Figure 11** presents the image reconstruction by elastic net method in ECT.

Figure 12 shows images of the experiments by ultrasound tomography. The algorithm was designed in such a way that an overdetermined system of equations could be generated, i.e., one for which the number of equations is greater than the number of unknowns. A feature of the tomography is, among other things, that the coefficient matrix is a rectangular deficiency of the pseudo-rank matrix. In such cases, you should consider trial solutions and choose only one of them. The obtained results are a raw tomographic image for synthetic data. In the numerical experiments presented, no additional adjustment method was used to obtain

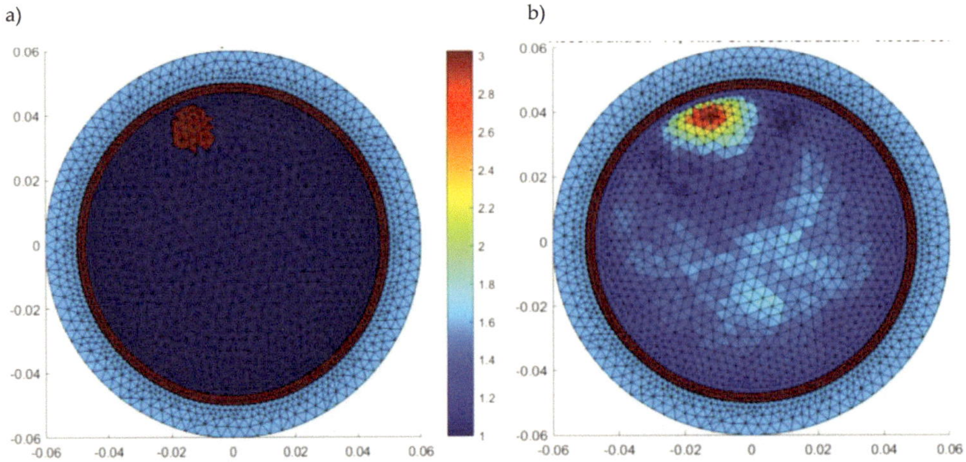

Figure 11. Image reconstruction by elastic net method (ECT): (a) model and (b) image reconstruction.

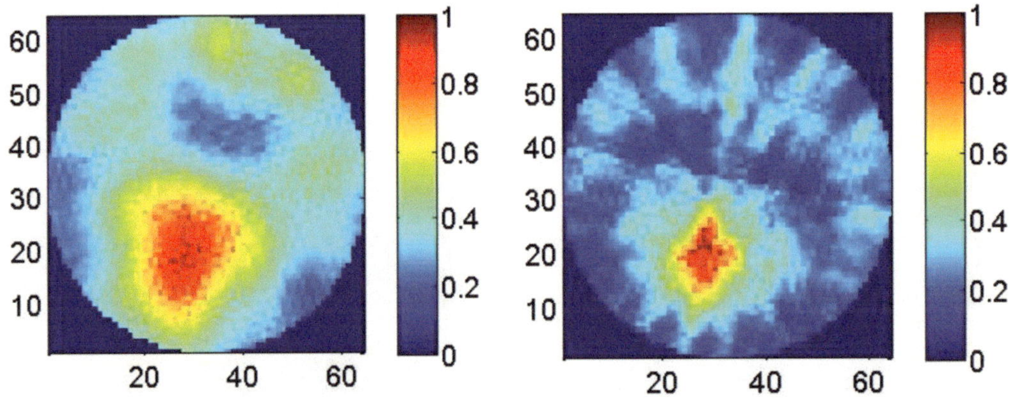

Figure 12. Localized on the 60 × 60 plane, tomographic images (UST).

images without streaks. The results obtained using the proposed method are a faithful representation of the modeled objects and enable their precise location in the considered area.

6. Conclusion

The chapter presents the idea of a control and optimization system based on tomographic sensors. Supervision and control belong to the scope of acquired and processed data and parameters of devices performing automation. Cyber-physical systems will change the way people interact and control the physical world. Industrial tomography can analyze physical and chemical phenomena without the need to penetrate into the interior. Electrical tomography and ultrasound tomography are included. The presented reconstructions were obtained by solving the inverse problem, which allows the imaging of processes. In the presented solution, we move from mathematical formalism to determining, analyzing, verifying, and checking systems that monitor and control physical processes. The system benefits various economic and industrial sectors. Advanced tools allow to capture both cybernetic abstractions and system dynamics. The system employs a communication interface, unique optimization algorithms, and data analysis algorithms for image reconstruction and process monitoring. The use of systems based on electrical and ultrasonic tomography can significantly improve industrial processes.

Author details

Tomasz Rymarczyk

Address all correspondence to: tomasz@rymarczyk.com

University of Economics and Innovation/Research and Development Center Netrix S.A, Lublin, Poland

References

[1] Hexmoor H, McLaughlan B, Tuli G. Natural human role in supervising complex control systems. Journal of Experimental and Theoretical Artificial Intelligence. 2009;**21**(1):59-77

[2] Kłosowski G, Kozłowski E, Gola A. Integer linear programming in optimization of waste after cutting in the furniture manufacturing. Advances in Intelligent Systems and Computing. 2018;**637**:260-270

[3] Ostrowski KL, Luke SP, Williams RA. Simulation of the performance of electrical capacitance tomography for measurement of de6nse phase pneumatic conveying. Chemical Engineering Journal. 1997;**68**:197-205

[4] Rybak G, Chaniecki Z, Grudzień K, Romanowski A, Sankowski D. Non–invasive methods of industrial process control, Informatyka, Automatyka. Pomiary w Gospodarce i Ochronie Środowiska. 2014;**4**(3):41-45

[5] Beck MS, Byars M, Dyakowski T, Waterfall R, He R, Wang SJ, Yang WQ. Principles and industrial applications of electrical capacitance tomography. Measurement and Control. September, 1997;**30**(7):197-200

[6] Garbaa H, Jackowska-Strumiłło L, Grudzień K, Romanowski A. Simulation of gravitational solids flow process and its parameters estimation by the use of electrical capacitance tomography and artificial neural networks. Informatyka, Automatyka, Pomiary w Gospodarce i Ochronie Środowiska. 2016;**6**(2):34-37

[7] Filipowicz SF, Rymarczyk T. The shape reconstruction of unknown objects for inverse problems. Przeglad Elektrotechniczny. 2012;**88**(3A):55-57

[8] Wang M. Industrial Tomography: Systems and Applications. Cambridge, UK: Elsevier; 2015

[9] Rymarczyk T, Tchórzewski P, Adamkiewicz P, Duda K, Szumowski J, Sikora J. Practical implementation of electrical tomography in a distributed system to examine the condition of objects. IEEE Sensors Journal. 2017;**17**(24):8166-8186. DOI: 10.1109/JSEN.2017.2746748

[10] Szulc K. Topological derivative – Theory and applications, Informatyka, Automatyka. Pomiary w Gospodarce i Ochronie Środowiska. 2015;**5**(1):7-13

[11] Szumowski J, Adamkiewicz P, Rymarczyk T. Electrical capacitance tomography to optimize quality control system. International Interdisciplinary Phd Workshop. 2016:121-124

[12] Voutilainen A, Lehikoinen A, Vauhkonen M, Kaipio JP. Three-dimensional nonstationary electrical impedance tomography with a single electrode layer. Measurement Science and Technology. 2010;**21**:035107

[13] Wajman R, Fiderek P, Fidos H, Jaworski T, Nowakowski J, Sankowski D, Banasiak R. Metrological evaluation of a 3D electrical capacitance tomography measurement system for two-phase flow fraction determination. Measurement Science and Technology. 2013;**24**:065302

[14] Lechleiter A, Rieder A. Newton regularizations for impedance tomography: Convergence by local injectivity. Inverse Problems. 2008;**24**(6):1-24

[15] Rymarczyk T. New methods to determine moisture areas by electrical impedance tomography. International Journal of Applied Electromagnetics and Mechanics. 2016; **37**(1-2):79-87

[16] Sbarbaro D, Vauhkonen M, Johansen TA. State estimation and inverse problems in electrical impedance tomography: Observability, convergence and regularization. Inverse Problems. 2015;**31**(3):045004

[17] Soleimani M, Mitchell CN, Banasiak R, Wajman R, Adler A. Four-dimensional electrical capacitance tomography imaging using experimental data. Progress In Electromagnetics Research. 2009;**90**:171-186

[18] Smolik W, Radomski D. Performance evaluation of the iterative image reconstruction algorithm with sensitivity matrix updating based on real measurements for electrical capacitance tomography. Measurement Science and Technology. 2009;**20**(11):115502

[19] Adler A, Lionheart WRB. Uses and abuses of EIDORS: An extensible software base for EIT. Physiological Measurement. 2006;**27**:25-42

[20] Holder DS. Electrical Impedance Tomography: Methods, History and Applications, Series in Medical Physics and Biomedical Engineering. 1st ed. Bristol, UK: Institute of Physical Publishing Ltd; 2005

[21] Ito K, Kunish K, Li Z. The level-set function approach to an inverse Interface problem. Inverse Problems. 2001;**17**(5):1225-1242

[22] Polakowski K, Filipowicz SF, Sikora J, Rymarczyk T. Quality of imaging in multipath tomography. Przeglad Elektrotechniczny. 2009;**85**(12):134-136

Electromechanical Co-Simulation for Ball Screw Feed Drive System

Liang Luo and Weimin Zhang

Additional information is available at the end of the chapter

http://dx.doi.org/10.5772/intechopen.80716

Abstract

Ball screw feed drive system is the most widely used linear drive system in the field of industrial automation. The continuous search for efficiency puts forward higher requests to the machine tool for high speed and high acceleration, which makes the feed drive system of lightweight-designed and large-size machine tools more likely to produce vibration during high-speed and high-acceleration feed operation. Electromechanical co-simulation for ball screw feed drive dynamics is an important technique for solving vibration problems occurring in the feed motion. This chapter elaborates on this technology from three aspects: modeling and simulation of dynamic characteristics of ball screw feed drive, modeling and simulation of servo control system, and the electromechanical co-simulation of ball screw feed drive system. In this chapter, the basic theoretical models, the establishment of simulation models and the comparison between simulation and experiment results of ball screw feed drive system are comprehensively introduced to provide technical references for readers.

Keywords: ball screw feed drive system, dynamic characteristics, electromechanical co-simulation, vibration, lumped mass model

1. Introduction

Ball screw feed system is the most widely used linear drive system in the field of industrial automation [1]. In order to enhance the speed and accuracy of present systems further, current research focuses on the vibration reduction and avoidance of the feed drive. Additional damping modules or structures are integrated in the feed drive system to achieve this goal, such as semi-active damping system, set point filtering, etc. Active damping system only reacts once a vibration is present, and set point filtering can lead to path deformation [2–4]. Another way to solve this problem is to generate a smoother trajectory. For this purpose, numbers of trajectory algorithms are found out, and the frequency contents of the trajectory orders are discussed and compared [5, 6]. The vibration caused by the trajectory is difficult to analyze on hardware because of the coupling factor of variety excitation sources. All these researches need a simulation method to help the researchers or engineers study or optimize the design and parameter setting of the feed drive system [10].

Finite element model of ball screw feed drive system can predict the accurate dynamic characteristics. However, it is difficult to integrate with the simulation model of servo control system. Lumped parameter model of ball screw feed drive system can simplify the simulation model by reducing the number of degrees of freedom (DOF) of the whole system. More importantly, it can easily integrate with the simulation model of the servo control system. A reasonable simplification of the lumped parameter model is the key to accurately predict the vibration of feed drive system [7–9].

In this chapter an electromechanical co-simulation method for ball screw feed drive system was established, which can be used to study the dynamic characteristics and vibration behavior of the feed drive system. An optimized dynamic modeling and simulation method of a ball screw feed drive based on the lumped mass model was firstly presented, and the optimized calculation method of the equivalent parameters was given. Then, a model of servo control system was built up, and based on it, the electromechanical co-simulation of ball screw feed drive system was established. Finally, a simulative and experimental test is conducted based on a ball screw feed drive system test bench. The result shows that electromechanical co-simulation of ball screw feed drive system could achieve a very good predictability.

2. Dynamic characteristic modeling and simulation of ball screw feed system

2.1. Lumped mass model of ball screw feed system

A typical ball screw feed system consists of a servomotor, coupling, ball screw, work table, and base (**Figure 1**). The ball screw is supported by two sets of bearing, which are fixed to the base. The servomotor torque is transmitted through a coupling onto the ball screw shaft to drive the work table. The linear guideway constrains the movement of the work table in an axial direction. The base is fixed on the machine bed or placed on the ground. The transformation from the rotational movement of the screw shaft into the linear motion of the work table is realized by the ball screw system with its transmission ratio i, which is defined as the distance of travel h during one revolution of the shaft as the following:

$$i = \frac{h}{2\pi} \tag{1}$$

Low-order modes are the main factors affecting the dynamic characteristics of the ball screw feed drive system of machine tools. Typically, the first axial and rotational modes of the ball screw show a dominant influence on the overall dynamics, while the relevance of higher-order modes for most technical applications is rather small [8].

The lumped mass model can reasonably reduce the number of degrees of freedom (DOF) of the simulation model while preserving the low-order modes of the system to simplify calculations. **Figure 2** shows the lumped mass model of a ball screw feed drive system. The influence of the shaft on the rotational mode and axial mode of the drive system is explicitly included into the lumped mass model here. Therefore, the shaft is separated into two different branches, an axial branch and a rotational branch, while the coupling once more is realized using constrained equations. Since all components are expressed by discrete springs and dampers, the rigidity values of shaft, coupling, and bearing are combined to an overall axial K_{ax} and rotational value K_{rot}.

In this model the inertial component parameters are defined as the following: rotary inertia of servomotor J_M, screw shaft side equivalent rotary inertia J_S, mass of base M_B, screw shaft side equivalent mass M_S, and mass of the work table M_T.

The equivalent rigidity parameters in the model are defined as the following: equivalent torsional rigidity K_{rot}, equivalent axial rigidity K_{ax}, rigidity of ball screw nut K_n, and axial rigidity of the base K_B.

The equivalent damping parameters are defined as the following: servomotor torsional damping C_M, equivalent torsional damping C_{rot}, screw shaft side damping C_S, ball screw nut damping C_n, equivalent axial damping C_{ax}, axial damping of the base C_B, and axial damping of the guide C_g.

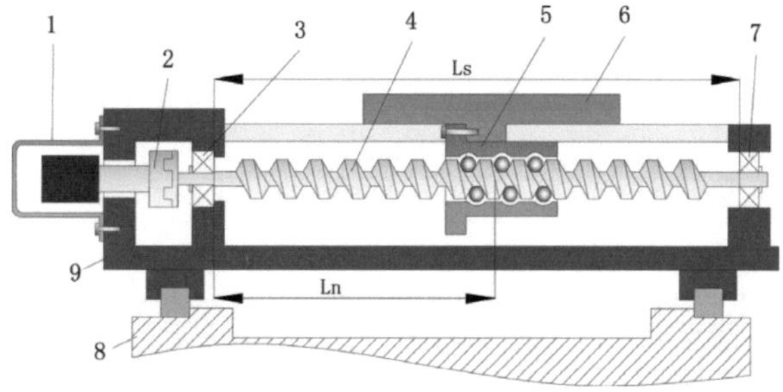

Figure 1. Typical structure of ball screw drive system. 1. Servomotor; 2. Coupling; 3. Fixed bearing; 4. Screw shaft; 5. Ball screw nut; 6. Work table; 7. Support bearing; 8. Machine bed; 9. Base.

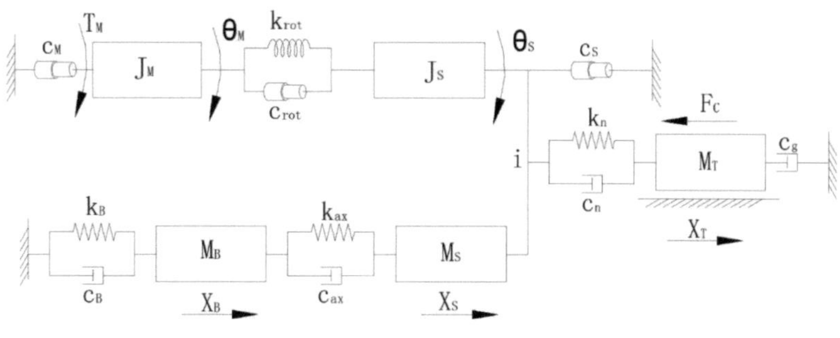

Figure 2. Lumped mass model of ball screw feed system.

The DOF parameters of the lumped mass model are defined as the following: angular rotation of the servomotor θ_M, screw shaft angular rotation at the table position θ_S, axial displacement of the base X_B, screw shaft axial displacement at the table position X_S, and work table position X_T. Therefore, the deformation of the equivalent springs is described as follows: equivalent torsional spring deformation $\theta_M - \theta_S$, equivalent axial spring deformation $X_S - X_B$, axial spring deformation of the base X_B, and screw nut contact deformation $X_T - X_S - i\theta_S$.

The speed parameters of equivalent damping are defined as the following: servomotor equivalent damping speed $\dot{\theta}_M$, equivalent torsional vibration damping speed $\dot{\theta}_M - \dot{\theta}_S$, equivalent damping speed of the screw $\dot{\theta}_S$, equivalent damping speed of the base $\dot{X}_B$, equivalent axial damping speed of screw $\dot{X}_S - \dot{X}_B$, equivalent damping speed of screw nut $\dot{X}_T - \dot{X}_S - i\dot{\theta}_S$, and speed of the work table $\dot{X}_T$.

According to the Lagrange's equations of the second kind, the dynamic model of the ball screw feed drive system is built up. The total kinetic energy T, the potential energy U, and the dissipation function of the system can be expressed using equations (2) through (4):

$$T = \frac{1}{2}J_M\dot{\theta}_M^2 + \frac{1}{2}J_S\dot{\theta}_S^2 + \frac{1}{2}M_B\dot{X}_B^2 + \frac{1}{2}M_S\dot{X}_S^2 + \frac{1}{2}M_T\dot{X}_T^2 \tag{2}$$

$$U = \frac{1}{2}k_{rot}(\theta_M - \theta_S)^2 + \frac{1}{2}k_B X_B^2 + \frac{1}{2}k_{ax}(X_S - X_B)^2 + \frac{1}{2}k_{nut}(X_T - X_S - i\theta_S)^2 \tag{3}$$

$$\begin{aligned} D = {} & \frac{1}{2}C_M\dot{\theta}_M{}^2 + \frac{1}{2}C_{rot}(\dot{\theta}_M - \dot{\theta}_S)^2 + \frac{1}{2}C_S\dot{\theta}_S{}^2 + \frac{1}{2}C_B\dot{X}_B{}^2 \\ & + \frac{1}{2}C_{ax}(\dot{X}_S - \dot{X}_B)^2 + \frac{1}{2}C_n(\dot{X}_T - \dot{X}_S - i\dot{\theta}_S)^2 + \frac{1}{2}C_T\dot{X}_T{}^2 \end{aligned} \tag{4}$$

According to the definition of the system lumped mass, we have the independent coordinates system $\mathbf{q}$ as the following:

$$\mathbf{q} = (\theta_M \;\; \theta_S \;\; X_B \;\; X_S \;\; X_T)^T \tag{5}$$

The force inputs of the ball screw feed system are the servomotor torque T_M and cutting force F_C, and then the generalized forces $\mathbf{Q}$ of the system can be expressed as the following:

$$\mathbf{Q} = (T_M \;\; 0 \;\; 0 \;\; 0 \;\; -F_C)^T \tag{6}$$

With $L = T - U$, the Lagrangian function of the system about the generalized coordinate $\mathbf{q}$ and the generalized force $\mathbf{Q}$ can be calculated according to Eq. (7). Then, the matrix form of the lumped mass model of the ball screw feed system can be established as in Eq. (8):

$$\frac{d}{dt}\left(\frac{\partial L}{\partial \dot{\mathbf{q}}}\right) - \frac{\partial L}{\partial \mathbf{q}} - \frac{\partial D}{\partial \dot{\mathbf{q}}} = \mathbf{Q} \tag{7}$$

$$\mathbf{m}\ddot{\mathbf{q}} + \mathbf{c}\dot{\mathbf{q}} + \mathbf{k}\mathbf{q} = \mathbf{Q} \tag{8}$$

where,

$$\mathbf{m} = \begin{bmatrix} J_M & 0 & 0 & 0 & 0 \\ 0 & J_S & 0 & 0 & 0 \\ 0 & 0 & M_B & 0 & 0 \\ 0 & 0 & 0 & M_S & 0 \\ 0 & 0 & 0 & 0 & M_T \end{bmatrix} \quad k = \begin{bmatrix} k_{rot} & -k_{rot} & 0 & 0 & 0 \\ -k_{rot} & k_{rot}\text{-}i^2k_n & 0 & -ik_n & ik_n \\ 0 & 0 & k_{ax}+k_B & -k_{ax} & 0 \\ 0 & ik_n & -k_{ax} & k_{ax}+k_n & -k_n \\ 0 & -ik_n & 0 & -k_n & k_n \end{bmatrix}$$

$$c = \begin{bmatrix} c_M + c_{rot} & -c_{rot} & 0 & 0 & 0 \\ -c_{rot} & c_{rot}\text{-}c_S\text{-}i^2c_n & 0 & -ic_n & ic_n \\ 0 & 0 & c_{ax}+c_B & -c_{ax} & 0 \\ 0 & ic_n & -c_{ax} & c_{ax}+c_n & -c_n \\ 0 & -ic_n & 0 & -c_n & c_n + c_g \end{bmatrix}$$

The dynamic model of the ball screw feed system shown in Eq. (8) was decomposed into three subsystems: screw shaft torsional vibration system, screw shaft axial vibration system, and the table vibration system. The simulation model of the ball screw feed system can be established as **Figure 3**. The input of the simulation model is the motor torque T_M and the cutting force F_C, and the outputs are the table acceleration a_T and displacement X_T.

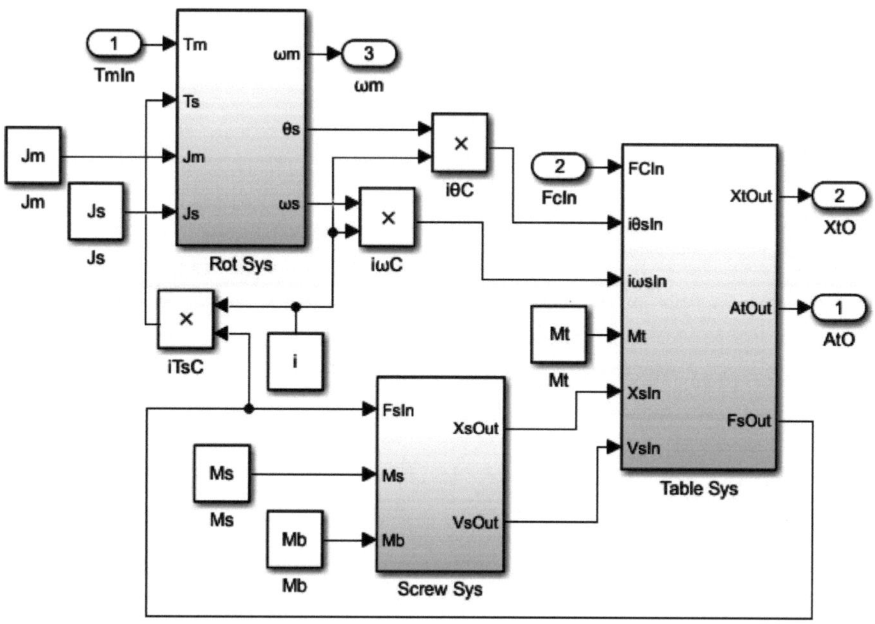

Figure 3. Simulation model of a ball screw feed drive.

2.2. Equivalent parameter calculation method of ball screw feed system lumped mass model

Accurate ball screw feed system dynamic model requires a reasonable equivalent parameter calculation method of the lumped mass model. As mentioned the shaft has influence on the rotational mode and the axial mode of the drive system; the shaft is separated into two different branches, an axial branch and a rotational branch, while the coupling is realized using constrained equations. The dynamic characteristics of the feed system should be analyzed to select the appropriate equivalent parameter calculation method. The inertia of the axial system and the inertia of rotational system are not only the mass or inertia of the component itself but also the mass or inertia converted to the independent coordinate system component of the dynamic system.

The equivalent rotary inertia of the screw J_S is composed of the rotary inertia of the screw J_{sc}, the rotary inertia of the coupling J_c, and the mass of the table M_T converted to the rotary inertia of the screw:

$$J_S = J_{sc} + J_c + M_T i^2 \tag{9}$$

With the material density ρ, namely, equivalent diameter d_s and length l_s of the screw shaft, the rotary inertia of the screw J_{sc} can be approximated using the following equation:

$$J_{sc} = \frac{\rho \cdot \pi}{32} d_s^4 l_s \tag{10}$$

The screw equivalent mass M_S is composed of screw mass M_{sc}, servomotor rotor mass M_m, and coupling mass M_c; with the material density ρ, namely, equivalent diameter d_s and length l_s of the screw shaft, the rotary inertia of the screw values J_{sc} can be approximated using the following equations:

$$M_S = M_{sc} + M_m + M_c \tag{11}$$

$$M_{sc} = \frac{\rho\pi}{4} d_s^2 l_s \tag{12}$$

The axial rigidity of the ball screw feed system is related to the installation method of the screw. Here is an example of the screw-fixed-support method used on most machine tools (**Figure 4**). Servomotor side of the screw shaft uses fixed support to provide screw axial support, and the end of the shaft is free support. Therefore, the axial rigidity k_{ax} of the ball

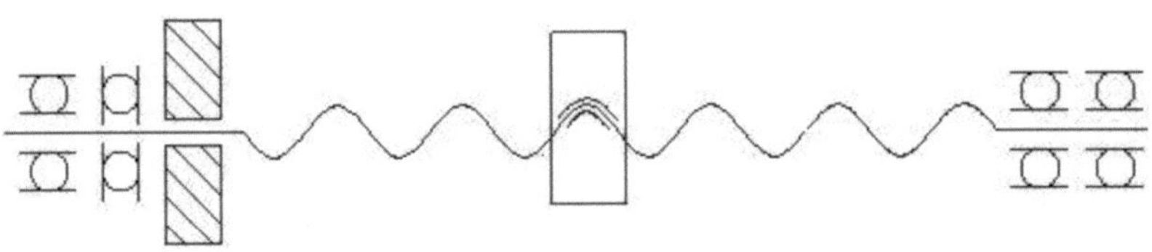

Figure 4. Screw fixation-support installation diagram.

screw feed system consists of screw bearing rigidity k_b, and the axial rigidity of the screw shaft $k_{s_{ax}}$ is as follows:

$$k_{ax} = \left(\frac{1}{k_b} + \frac{1}{k_{s_{ax}}}\right)^{-1} \tag{13}$$

With the material elastic modulus E, screw cross-sectional area A, and ball screw length at table position l_n, the axial rigidity of the screw according to the position of the table $k_{s_{ax}}$ can be established as shown in Eq. (14):

$$k_{s_{ax}} = \frac{EA}{l_n} = \frac{\pi d_s^2 E}{4 l_n} \tag{14}$$

The torsional rigidity of the ball screw feed system k_{rot} consists of the torsional rigidity of the screw $k_{s_{rot}}$ and the torsional rigidity of the coupling k_c. According to the position of the table, the torsional rigidity of the screw $k_{s_{rot}}$ can be approximated with the shear modulus G and polar rotary inertia of the screw section I_P as shown in Eq. (16):

$$k_{rot} = \left(\frac{1}{k_c} + \frac{1}{k_{s_{rot}}}\right)^{-1} \tag{15}$$

$$k_{s_{rot}} = \frac{T}{\Delta\theta} = \frac{G \cdot I_P}{l_n} = \frac{G \cdot \pi \cdot d_s^4}{32 \cdot l_n} \tag{16}$$

With the nut reference rigidity K, nut axial load F_a, and basic dynamic load C_a, the contact rigidity of the screw nut k_n can be expressed as follows:

$$k_n = 0.8K\left(\frac{F_a}{0.1C_a}\right)^{\frac{1}{3}} \tag{17}$$

3. Servo control system modeling and simulation

3.1. Modeling of permanent magnet synchronous motor

Permanent magnet synchronous motors can be divided into two types according to the rotor type, salient pole rotor and non-salient pole rotor. The structure is shown in **Figure 5**; in the surface-mounted permanent magnet synchronous motor (**Figure 5(a)**), the magnetic circuit of the rotor is symmetrical, and the magnetic permeability and air gap permeability of the permanent magnet material are approximately the same. In the rotor two-phase coordinate system, the direct-axis inductance and the quadrature-axis inductance are equal, that is $L_d = L_q$. It is named non-salient pole rotor permanent magnet synchronous motor.

The rotor magnetic paths of plug-in-type (**Figure 5(b)**) and built-in-type (**Figure 5(c)**) permanent magnet synchronous motors are asymmetrical, and the quadrature-axis inductance is

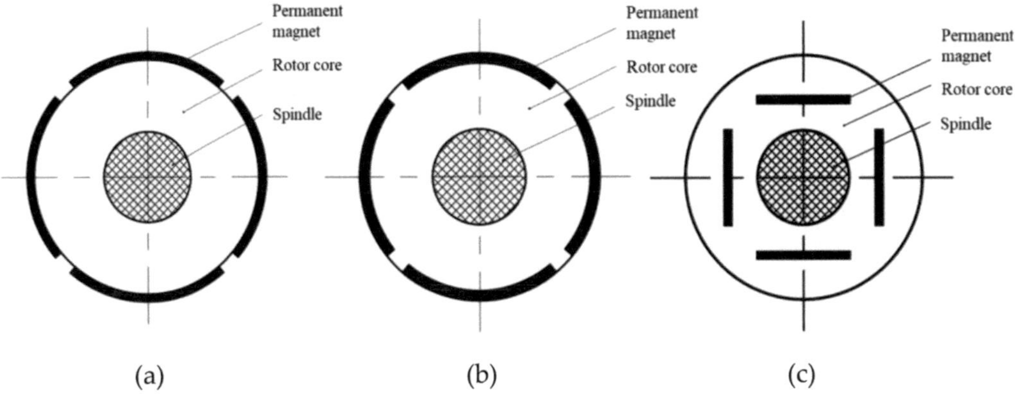

Figure 5. PMSM motor rotor structure. (a) Surface-mounted; (b) Plug-in; and (c) Interior-mounted.

greater than the direct-axis inductance, that is $L_q > L_d$. The rotor shows salient pole effect, which is called salient pole-type permanent magnet synchronous motor.

Taking non-salient pole rotor permanent magnet synchronous motor as an example, we simplify the motor model with the following conditions: neglecting the saturation of the motor core; no eddy current and hysteresis loss; permanent magnet material has zero conductivity; three-phase windings are symmetrical; and induced potential in the winding is sinusoidal. Then, a schematic diagram of the physical model of the motor shown in **Figure 6** can be obtained.

The axis of the sinusoidal magnetomotive wave generated by a flowing forward current through the phase winding is defined as the axis of the phase winding. Take axis A as the spatial reference coordinate of the ABC coordinate system. It is assumed that the positive direction of the induced electromotive force is opposite to the positive direction of the current (motor principle); take the counterclockwise direction as the positive direction of the speed and

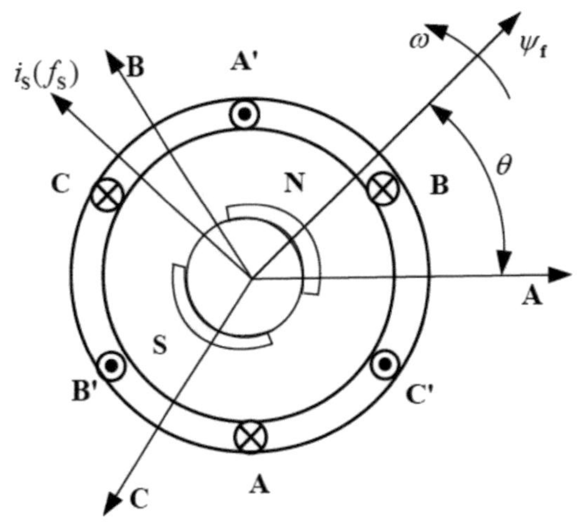

Figure 6. PMSM physical model.

electromagnetic torque, and the positive direction of the load torque is the opposite. The physical model equation of permanent magnet synchronous motor is as follows:

$$\begin{cases} U_S = Ri_s + \dfrac{d\Psi_s}{dt} \\ \Psi_s = Li_s + \Psi_f \begin{bmatrix} \cos(\theta + \omega t) \\ \cos(\theta + \omega t + 2\pi/3) \\ \cos(\theta + \omega t - 2\pi/3) \end{bmatrix} \end{cases} \tag{18}$$

$$\begin{bmatrix} u_A \\ u_B \\ u_C \end{bmatrix} = \begin{bmatrix} R & 0 & 0 \\ 0 & R & 0 \\ 0 & 0 & R \end{bmatrix} \cdot \begin{bmatrix} i_A \\ i_B \\ i_C \end{bmatrix} + \begin{bmatrix} \dfrac{d\varphi_A(\theta, i)}{dt} \\ \dfrac{d\varphi_A(\theta, i)}{dt} \\ \dfrac{d\varphi_A(\theta, i)}{dt} \end{bmatrix} \tag{19}$$

In Formula (18), R and L are matrices and their expressions are shown in Formula (20):

$$\begin{cases} U_s = [U_A \quad U_B \quad U_C]^T \\ i_s = [i_A \quad i_B \quad i_C]^T \\ R = \begin{bmatrix} R_S & 0 & 0 \\ 0 & R_S & 0 \\ 0 & 0 & R_S \end{bmatrix} \\ L = \begin{bmatrix} L_A & M_{AB} & M_{AC} \\ M_{BA} & L_B & M_{BC} \\ M_{CA} & M_{CB} & L_C \end{bmatrix} \end{cases} \tag{20}$$

U_A, U_B, and U_C are phase voltages of PMSM stator three-phase windings. I_A, I_B, and I_C are the phase currents of stator three-phase windings. R_S is the stator winding value. L_A, L_B, and L_C are stator winding self-inductances. $M_{AB} = M_{BA}$, $M_{AC} = M_{CA}$, and $M_{BC} = M_{CB}$ are the stator winding mutual inductances.

3.2. Coordinate transformation

From the above physical model, we can see that in the ABC coordinate system, the PMSM rotor is asymmetric in the magnetic and electrical structures. The motor equation is a set of nonlinear time-varying equations related to the instantaneous position of the rotor, which makes the analysis of the dynamic characteristics of the PMSM very difficult. It is usually necessary to convert the motor equations by coordinate transformation to facilitate analysis and calculation. The coordinate system used in the vector control of the permanent magnet synchronous motor and their relationship is shown in **Figure 7**. In the figure, the $\alpha - \beta$

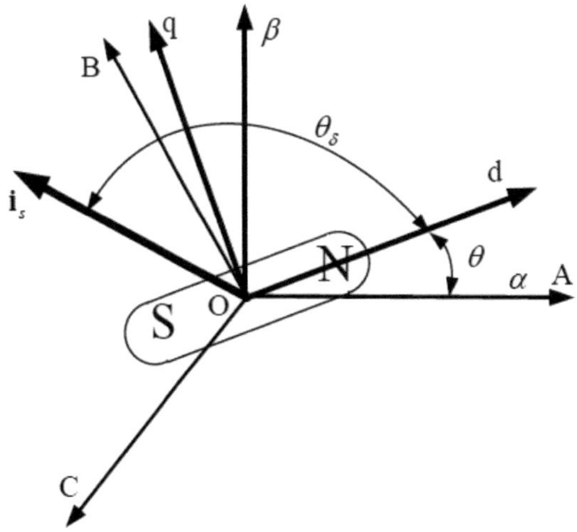

Figure 7. The coordinate system used in the vector control and their relationship.

coordinate system is a two-phase stationary coordinate system, and the $d-q$ coordinate system is a two-phase rotating coordinate system that is fixed to the rotor.

3.2.1. Clarke transformation

Clarke transformation simplifies the voltage loop equations on the original three-phase windings into the voltage loop equations on the two-phase windings, from the three-phase stator ABC coordinate system to the two-phase stator $\alpha-\beta$ coordinate system, as shown in Eq. (21); its inverse transform is shown in Eq. (22). However, after the Clarke transformation, the torque still depends on the rotor flux. In order to facilitate control and calculation, Park transformation is also required:

$$\begin{bmatrix} i_\alpha \\ i_\beta \end{bmatrix} = \sqrt{\frac{2}{3}} \begin{bmatrix} 1 & -1/2 & -1/2 \\ 0 & \sqrt{3}/2 & -\sqrt{3}/2 \end{bmatrix} \begin{bmatrix} i_A \\ i_B \\ i_C \end{bmatrix} \tag{21}$$

$$\begin{bmatrix} i_A \\ i_B \\ i_C \end{bmatrix} = \sqrt{\frac{3}{2}} \begin{bmatrix} 1 & 0 \\ -1/2 & \sqrt{3}/2 \\ -1/2 & -\sqrt{3}/2 \end{bmatrix} \begin{bmatrix} i_\alpha \\ i_\beta \end{bmatrix} \tag{22}$$

3.2.2. Park transformation

In the physical sense, the Park transformation is equivalent to projecting the currents i_a, i_b, i_c onto the $d-q$ axis. The transformed coordinate system rotates at the same speed as the rotor, and the d-axis and the rotor flux have the same position. The Park transformation is shown in Eq. (23), and the inverse transformation is shown in Eq. (24). This transformation also holds for the three-phase voltage and flux linkage:

$$\begin{bmatrix} i_d \\ i_q \end{bmatrix} = \begin{bmatrix} \cos\theta & \sin\theta \\ -\sin\theta & \cos\theta \end{bmatrix} \begin{bmatrix} i_\alpha \\ i_\beta \end{bmatrix} \tag{23}$$

$$\begin{bmatrix} i_\alpha \\ i_\beta \end{bmatrix} = \begin{bmatrix} \cos\theta & -\sin\theta \\ \sin\theta & \cos\theta \end{bmatrix} \begin{bmatrix} i_d \\ i_q \end{bmatrix} \tag{24}$$

3.2.3. *Mathematical model of permanent magnet synchronous motor in coordinate system*

The mathematical model of the permanent magnet synchronous motor in the ABC coordinate system can be transformed into any two-phase coordinate system through coordinate transformation, so that it is possible to simplify the decoupling of the motor flux linkage equation and the electromagnetic torque equation. If the mathematical model of the motor is transformed into a $d-q$ coordinate system fixed on a permanent magnet rotor, the motor flux equation and the electromagnetic torque equation will be greatly simplified.

With the equivalent flux ψ_d and ψ_q, the equivalent inductance L_d, L_q of the motor in the $d-q$ coordinate system, and the rotor permanent magnet flux linkage ψ_f, the stator flux equation can be obtained in Eq. (25):

$$\begin{cases} \psi_d = L_d i_d + \psi_f \\ \psi_q = L_q i_q \end{cases} \tag{25}$$

Take the rotor permanent magnet flux linkage ψ_f as constant; the voltage of stator in the $d-q$ coordinate system is as follows:

$$\begin{cases} u_d = Ri_d + L_d \dfrac{di_d}{dt} - \omega_r L_q i_q \\ u_q = Ri_q + L_q \dfrac{di_q}{dt} + \omega_r L_d i_d + \omega_r \psi_f \end{cases} \tag{26}$$

where u_d and u_q are the stator-side equivalent voltages of the motor in the $d-q$ coordinate system, R is the stator winding resistance per phase, and ω_r is the electrical angular velocity of the rotor rotation.

The electromagnetic torque equation in the $d-q$ coordinate system can be expressed as follows, where p is the number of pole pairs, $\boldsymbol{\psi}$ is the flux synthesis vector, and $\mathbf{i}$ is the current composition vector:

$$T_e = 1.5p\boldsymbol{\psi} \times \mathbf{i} \tag{27}$$

Using the components of the $d-q$ coordinate system to represent the flux linkage and current vector shown in Eq. (28), the electromagnetic torque can be expressed as shown in Eq. (29):

$$\begin{cases} \mathbf{i} = i_d + ji_q \\ \boldsymbol{\psi} = \psi_d + j\psi_q \end{cases} \tag{28}$$

$$T_e = 1.5p\left[\psi_f i_q + \left(L_d - L_q\right) i_d i_q\right] \tag{29}$$

3.3. Servo control modeling of ball screw feed system

In order to model the servo control of the ball screw feed system, the modeling of the three-loop cascade control architecture of the vector control and servo control system of the permanent magnet synchronous servomotor is studied, which is commonly used in the ball screw feed system.

3.3.1. Modeling of permanent magnet synchronous motor vector control.

The principle of space vector pulse width modulation (SVPWM) is based on vector equivalents. The magnitude and direction of the current vector can be indirectly controlled by the timing of the six switching elements of the inverter through the three-phase winding of the permanent magnet synchronous motor, so that the winding produces a constant amplitude circular magnetic field that rotates according to a given demand, thus dragging the permanent magnet to rotate. The voltage inverter circuit is shown in **Figure 8**. The simulation model of SVPWM control system for permanent magnet synchronous motor is shown in **Figure 9**.

3.3.2. Modeling of cascade control system

The SVPWM control system for permanent magnet synchronous motor is based on the three-phase current information and rotor position information fed back by the motor. AC motor is equivalent to a direct current motor by formula transformation to control the position and amplitude of the stator current.

The control system schematic is shown in **Figure 10(a)**. The system includes a cascaded control structure with a P-position controller, a PI-velocity controller, and a PI-current controller. In the cascade control system, the servomotor feedback speed ω_M and the work table feedback

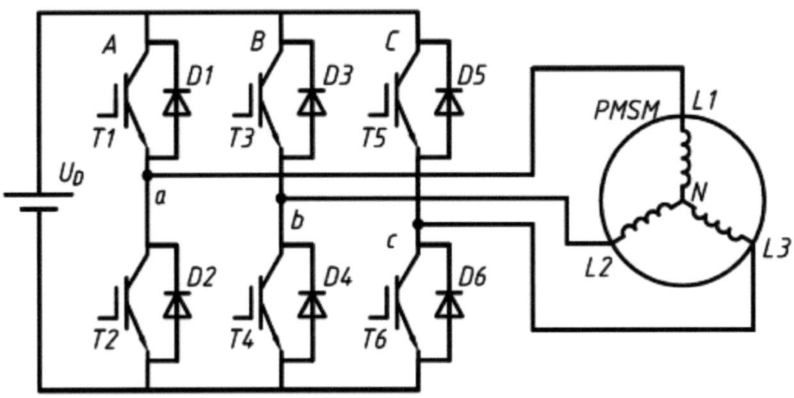

Figure 8. Circuit of voltage bridge inverter.

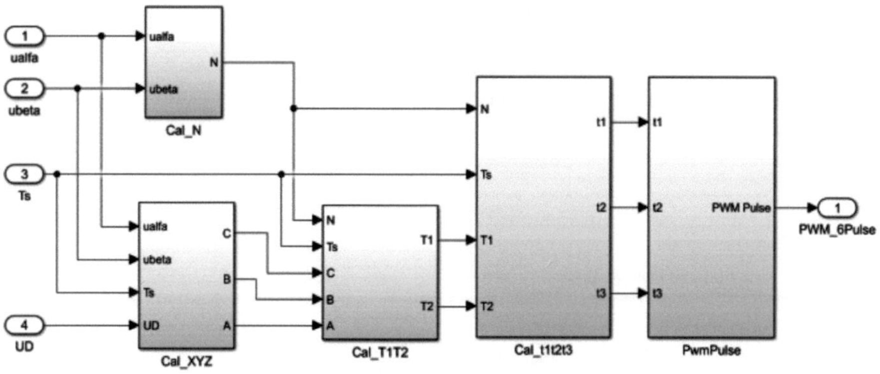

Figure 9. Simulation model of SVPWM control system.

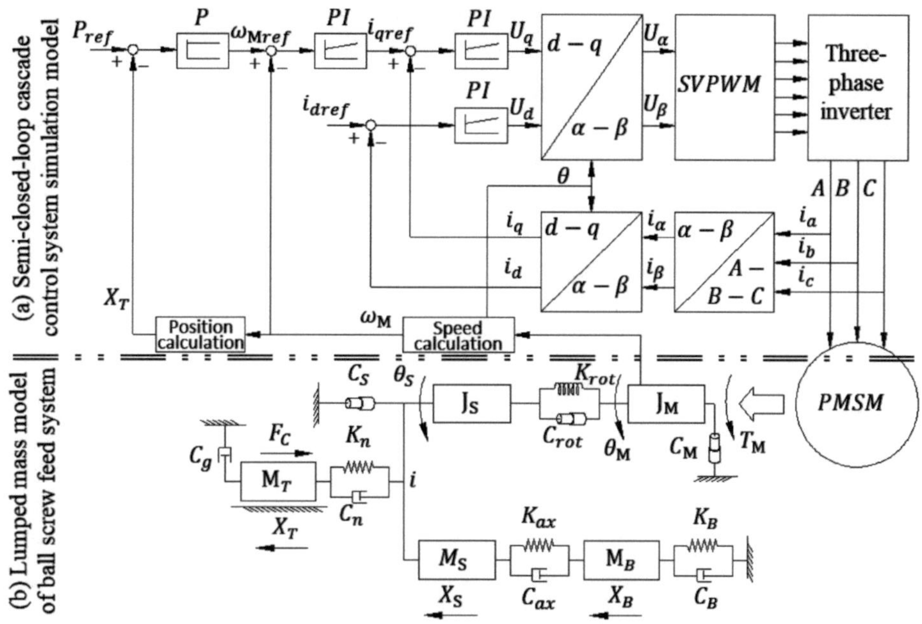

Figure 10. Schematic of ball screw feed drive system electromechanical co-simulation.

position X_T are calculated by the rotor position detected by the encoder on the servomotor. The input P_{ref} is given by the CNC system according to the feed motion command, the input P_{ref} and work table feedback position X_T are compared, and the reference speed ω_{Mref} is given by the position controller. Then, the reference speed ω_{Mref} is compared with the feedback speed ω_M, and the velocity controller gives the reference current i_{qref} for the q-axis and the reference current $i_{dref} = 0$ for the d-axis of the stator. The three-phase current of the servomotor is detected and converted into i_d and i_q in $d - q$ coordinate system through the Clark and Park

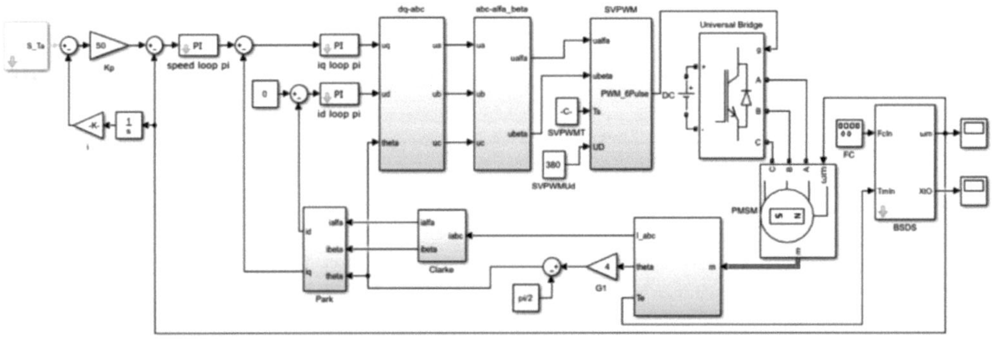

Figure 11. Electromechanical co-simulation model of half-closed ball screw feed system.

transformation. i_{dref} and i_{qref} are compared with the feedback i_d and i_q, respectively, and the current controller calculates the given voltages U_d and U_q of the d and q axes; then, they are converted into U_α and U_β in the $\alpha - \beta$ coordinate system by Park inverse transformation. Finally, the SVPWM module generates six-phase PWM to drive the three-phase inverter. The inverter outputs ABC three-phase voltage to servomotor stator, which generates rotating magnetic field and produces magnetic torque on the servomotor rotor. This magnetic torque is the output torque T_M of the servomotor and drives the rotor to rotate under the dynamic relations of ball screw feed system.

3.4. Electromechanical co-simulation modeling of ball screw feed drive system

Based on the lumped mass model of ball screw feed system and the servo control system simulation model, an electromechanical co-simulation model of the ball screw feed drive system was constructed. The co-simulation schematic is shown in **Figure 10**; as described above (a) is the semi-closed-loop cascade control system simulation model, while (b) is the lumped mass model of ball screw feed system. The inverter outputs ABC three-phase voltage to servomotor stator, which generates rotating magnetic field and produces magnetic torque on the servomotor rotor. This magnetic torque is the output torque T_M of the servomotor and drives the rotor to rotate under the dynamic relations as shown in Eq. (8).

The electromechanical co-simulation model of the ball screw feed drive system is shown in **Figure 11**. The S_Cal module on the left side generates the trajectory command for the feed drive system according to the acceleration/deceleration strategy. Under the cascade control system, which consists of position controller, velocity controller, and current controller, the servomotor drive and the ball screw accomplish the motion command accordingly.

4. Experimental verification of the electromechanical co-simulation model of ball screw feed drive system

The electromechanical co-simulation model in this chapter has been tested on a single-axis ball screw drive system test bench shown in **Figure 12**. The test bench uses an i5 CNC system and

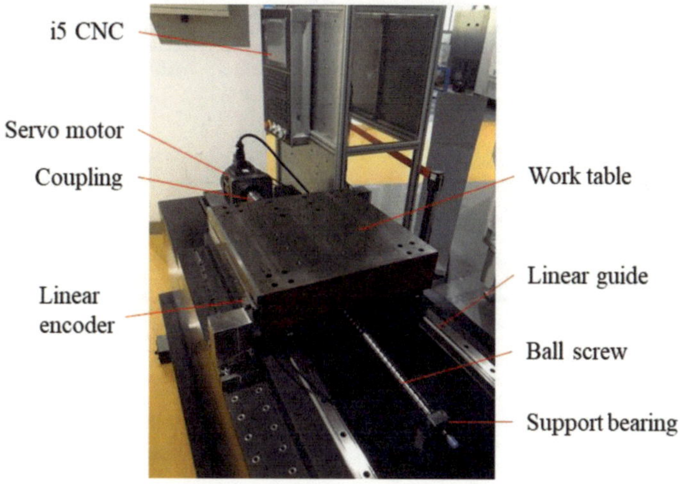

Figure 12. Single-axis ball screw feed drive test bench.

servo system of Shenyang Machine Group, which use a semi-closed-loop cascade control structure. The specifications of the test bench are listed in **Table 1**, which are either obtained from the manufacturers' catalogs, approximated from prior knowledge, or calculated from computer-aided design (CAD). According to the modeling method described in Chapter 2, the lumped mass model of this ball screw feed system test bench was built up. The equivalent parameters of the lumped mass model were calculated by using the specifications in **Table 1**, and the other calculated lumped parameters are listed in **Table 2**.

Taking the servomotor torque as input and the axial acceleration of work table as output, the frequency response characteristics of the lumped parameter model of the test bench are analyzed. The bode diagram is shown in **Figure 13**, and simulation result shows that the work table has four-order natural frequencies, which are 26.2, 76.7, 247, and 633 Hz. Further study shows that 76.7 Hz is the main axial vibration frequency of the work table, 26.2 Hz is the main axial vibration frequency of the base, and 247 and 633 Hz are the rotational vibration frequencies.

Parameter of the component	Value	Parameter of the component	Value
Work table mass M_T (kg)	206	Rotary inertia of coupling J_C $(kg \cdot m^2)$	1.09×10^{-4}
Base mass M_B (kg)	3820	Rotary inertia of motor J_M $(kg \cdot m^2)$	6.75×10^{-3}
Coupling mass M_c (kg)	1.18	Torsional rigidity of coupling k_c (N/m)	1.4×10^3
Motor rotor mass M_m (kg)	10.9	Screw bearing rigidity k_b (N/m)	1×10^8
Screw pitch length h (m)	1.6×10^{-2}	Nut reference rigidity K (N/m)	6.12×10^8
Screw diameter d_s (m)	2.5×10^{-2}	Nut basic dynamic load C_a (N)	37.4
Screw length l_s (m)	1	Ball screw length at table position l_n (m)	0.35

Table 1. Specifications of the test bench.

Parameter of the component	Value	Parameter of the component	Value
Screw equivalent mass M_S (kg)	11.28	Equivalent rotary inertia of screw J_S $(kg \cdot m^2)$	1.7×10^{-3}
Axial rigidity of screw K_{ax} (N/m)	0.743×10^8	Rotary rigidity of screw K_{rot} $(N \cdot m \cdot rad^{-1})$	3.14×10^3
Axial rigidity of base K_B (N/m)	1×10^8	Contact rigidity of the screw nut K_n (N/m)	9.8×10^7

Table 2. Calculated parameters used in the lumped mass model of test bench.

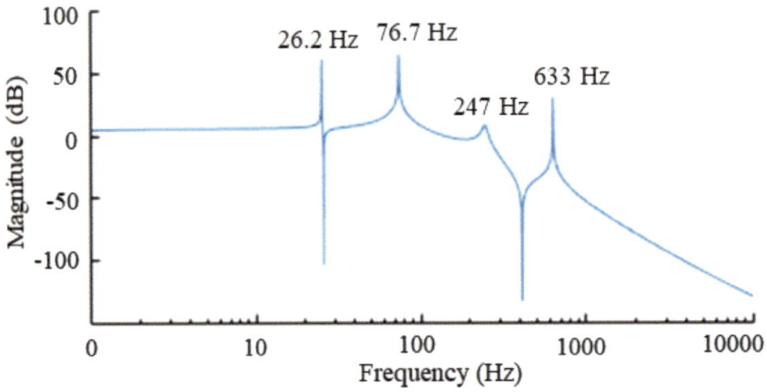

Figure 13. Bode diagram of the lumped parameter model of ball screw feed system.

To establish the simulation model of servo control system, the servomotor parameters are needed as shown in **Table 3**.

In order to compare and verify the simulation results with the experimental results, the motion command parameters and the control parameters of the experimental test and simulation are set in **Table 4**.

Parameter name	Value	Parameter name	Value
Rated power (kW)	4.4	Number of pole pairs	4
Rated torque $(N \cdot m)$	18.6	Stator resistance per phase (Ω)	1.44
Rotor inertia $(kg \cdot m^2)$	6.75×10^{-3}	Inductance L_d, L_q (H)	8.15×10^{-3}
Rated speed (r/min)	1500	Permanent magnetic flux $\psi_f(wb)$	0.21

Table 3. Parameters of the servomotor.

Parameter name	Value	Parameter name	Value
Position instruction (mm)	400	Position loop gain k_p	50
Maximum velocity (mm/s)	400	Velocity loop gain k_v	10
Maximum acceleration (mm/s^2)	2000	Current loop gain k_i	30
Maximum jerk (mm/s^3)	20,000		

Table 4. Motion command parameters and the control parameters settings.

Using the same operation parameters as set in the simulation model, a feed motion experiment was conducted and the work table position was measured. The simulation results are compared to the experimental results. **Figures 14** and **15** exemplarily show simulated and measured

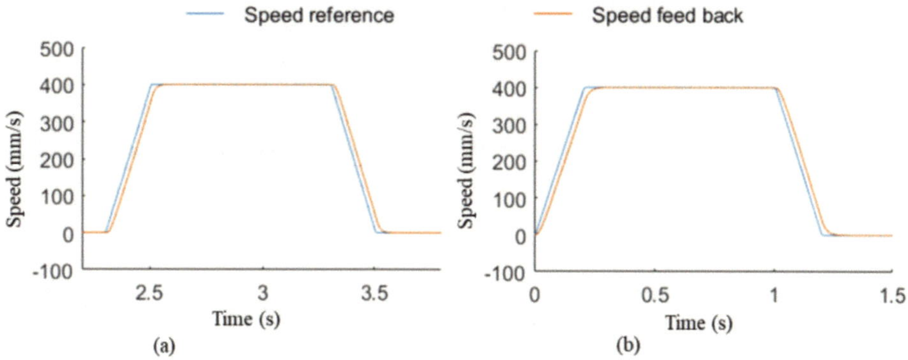

Figure 14. Reference velocity and feedback velocity.

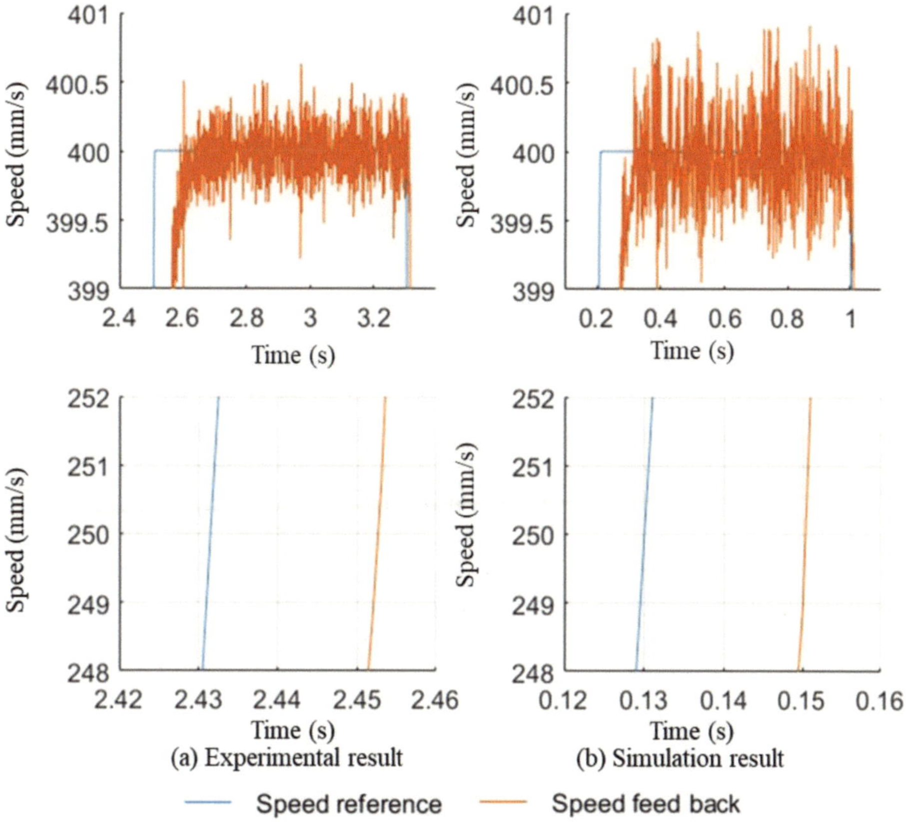

Figure 15. Detailed reference velocity and feedback velocity.

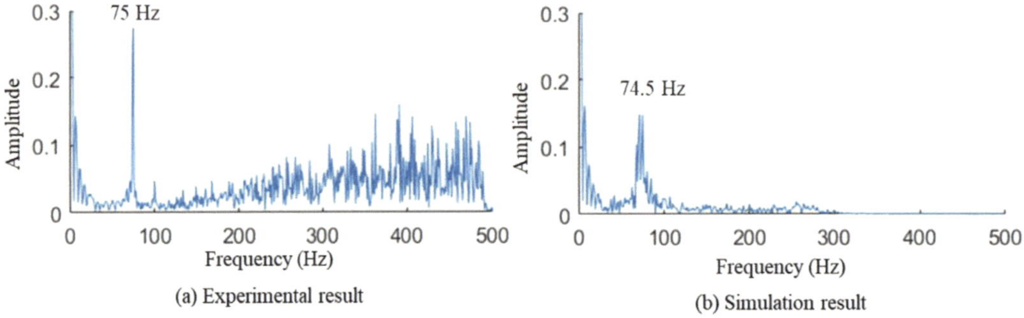

Figure 16. Frequency contents of work table acceleration.

reference velocity and feedback velocity of the servomotor at the given operating conditions. The simulation result has a similar curve to the experimental result.

Figure 16 shows frequency contents of the work table acceleration signals from simulation result and the experimental result. Comparing the simulation result with the experimental result, the co-simulation model of ball screw feed drive system can predict the vibration that occurs in the feed operation. Both results show that in this case the second-order natural frequency (about 75 Hz) but not the first-order natural frequency is the main factor influencing the performance of feed drive system.

5. Conclusions

In this chapter an electromechanical co-simulation model of the ball screw feed drive system was constructed based on lumped mass model of ball screw feed system and the servo control system simulation model, which can be used to study the dynamic characteristics and vibration behavior of the feed drive system. Simulative and experimental tests were conducted based on a ball screw feed drive system test bench. The result shows that the co-simulation model of ball screw feed drive system can predict the vibration that occurs in the feed operation. Because of the integration of lumped parameter model into the detailed modeled cascade control simulation model, the electromechanical co-simulation of ball screw feed drive system could achieve a very good predictability for control performance and vibration behavior study of ball screw feed drive system, which may be affected by the servo controller, ball screw feed system, or the coupling between them.

Acknowledgements

This work is supported by the major national science and technology projects "high-end CNC machine tools and basic manufacturing equipment" (2012ZX04005031).

Author details

Liang Luo and Weimin Zhang*

*Address all correspondence to: iamt.tongji.edu.cn

School of Mechanical Engineering, Tongji University, Shanghai, China

References

[1] Altintas Y, Verl A, et al. Machine tool feed drives. CIRP Annals–Manufacturing Technology. 2011;**60**:779-796

[2] Verl A, Frey S. Improvement of feed drive dynamics by means of semi-active damping. CIRP Annals–Manufacturing Technology. 2012;**61**:351-354

[3] Dietmair A, Verl A. Drive based vibration reduction for production machines. Science Journal. 2009, Oct:130-134

[4] Ming Y, Hao H, Dianguo X. Cause and suppression of mechanical resonance in PMSM servo system. Electric Machines and Control. 2012, Jan;**16**(1):79-84

[5] Jian-Ren S. Research on ACC/DEC Control and Contour Error of CNC System. Lan Zhou: Lanzhou University of Technology; 2012

[6] Erkorkmaz K, Altintas Y. High speed CNC system design. Part I: Jerk limited trajectory generation and quintic spline interpolation. International Journal of Machine Tools and Manufacture. 2001;**41**(9):1323-1345

[7] Sato R. Development of a feed drive simulator. Key Engineering Materials. 2012;**516**: 154-159

[8] Frey S, Dadalau A, Verl A. Expedient modeling of ball screw feed drives. Production Engineering. 2012;**6**(2):205-211

[9] Okwudire EC, Altintas Y. Hybrid modeling of ball screw drives with coupled axial, torsional and lateral dynamics. Journal of Mechanical Design. 2009;**131**:071002-1-071002-9

[10] Liang L, Weimin Z, Mingjian Z, et al. Dynamics modeling and simulation of ball screw feed drive based on lumped mass model. Transactions of the Chinese Society for Agricultural Machinery. 2015;**46**(12):370-377

Chapter 5

Industry 3.0 to Industry 4.0: Exploring the Transition

Shane Loughlin

Additional information is available at the end of the chapter

http://dx.doi.org/10.5772/intechopen.80347

Abstract

This work is a *How-To-Guide for Digit**ALIZA**tion of Industry 4.0 Manufacturing*. It provides a novel ALIZA Canvas and ALIZA Process supported by a comprehensive ALIZA Toolset. This output is derived from observed, tangible deficiencies in contemporary functional communications in manufacturing. This study proposes an innovative approach with robust methodologies for strategic alignment of the technical and business components in manufacturing. The requirement for a supplementary educational infrastructure, to address the pronounced educational shortcomings and knowledge gaps in the transition to Industry 4.0 is outlined. An explanation is provided of how E-Cubers (our own educational organization) will design, develop, and deliver educational programmes on Topics relevant to achieving Industry 4.0 Equipment Engineering Excellence. It defines and tests the novel concept of the E-Cubers Eight Ps; encompassing **p**rioritized **p**roblem solving, via **p**ortfolios and **p**rojects, through **p**eer collaboration within a defined technology playground with emphasis on learning and **p**laying with **p**assion. The E-Cubers Eight Ps is combined with *The E-Cubers Library* to deliver a truly comprehensive specialist, national learning framework. This holistic approach will ultimately enable Ireland to lead the way in Industry 4.0 by doing what we do best *"ag spraoi agus ag imirt" (Gaelic – playing by having fun and competing).*

Keywords: Industry 4.0 equipment, OEE, ALIZA, E-Cubers, LEGO, PBWS, OSE, DIVOM

1. Introduction

The goal of Industry 4.0 is *"The Intelligent Factory"*, which is characterized by adaptability, resource efficiency and ergonomics as well as the integration of customers and business partners in business and value processes. This *"Factory of the Future"*, will regard everything as a service (including tools and skills), it will require supply-chain integration and data availability. These combined elements will allow for the integration of the entire production process and supply chain and should, in time, enable the self-optimization of cyber-based physical systems [1].

Irish industry is currently poised for the transition from conventional automation of Industry 3.0 (I3) to the Cyber Physical Systems (CPS) of Industry 4.0 (I4). What will the transition look like? Where are the tools to manage the transition? In the absence of any definitive *How-To-Guides* business leaders depend heavily on advice from technology providers. This has the inherent risk of implementing overly technical solutions which may deliver very little business value. The primary focus for Irish businesses must be DigitALIZAtion which Gartner [2] defines as *the use of digital technologies to change a business model and provide new revenue and value-producing opportunities; it is the process of moving to a digital business*. This chapter is the *How-To-Guide for the DigitALIZAtion of I4 Manufacturing.*

Creating the *How-To-Guide for the DigitALIZAtion of I4 Manufacturing* required the research, development and validation of many novel processes and tools such as *The ALIZA Canvas, The ALIZA Process* and *The ALIZA Tools (I4-PS Scorecard, I4-ES Scorecard, The OSE Calculator and The DIVOM Benchmarking Process).* Each of these processes and tools are explained in detail in this chapter.

For tangible business benefit is to be derived from these processes and tools they must be rapidly disseminated to practitioners. This work proposes that there is a requirement to supplement the existing education function of discipline centric qualifications with Topic centric competencies which enable the Trans-Topic collaboration necessary for the creation of novel technical solutions to the emerging Industry 4.0 business problems. This requires a supplementary educational organization, E-Cubers, which consists of a constellation of Communities of Practice (CoPs) organized around topics which are designed to facilitate collaboration and creativity for the advancement of each members individual competencies to support the achievement of I4 Equipment Engineering Excellence.

A detailed explanation is provided of how E-Cubers have utilized LEGO® TECHNIC, LEGO® MINDSTORMS and LEGO® GBC as equipment kits to deliver the Overall Equipment Effectiveness (OEE) centric BUILD, PROGRAM and INVENT methods. A detailed evaluation of Resnick's Four Ps leads to the definition of E-Cubers Eight Ps as: *E-Cubers utilize the concept of an OEE* ***Playground*** *to* ***Passionately*** *build up a* ***Portfolio*** *of* ***Projects*** *in collaboration with their* ***Peers****, "ag spraoi agus ag imirt" (Gaelic for* ***Playing*** *by having fun and competing). They demonstrate their technical ability by leveraging existing knowledge assets and creating novel solutions to the* ***Problems*** *which they have* ***Prioritized*** *as they learn the art of their craft.*

But the work does not stop there. By creating *The E-Cubers Library* it becomes possible for E-Cubers to share their physical assets as efficiently as they utilize the Creative Commons Licensing model to share their cyber assets and knowledge. This is a truly comprehensive *How-To-Guide for the DigitALIZAtion of I4 Equipment Knowledge* which when fully harnessed will enable Ireland to take its rightful place as leaders in I4 Equipment by doing what we do best *"ag spraoi agus ag imirt."*

2. The DigitALIZAtion of I4 manufacturing

2.1. Introduction

DigitALIZAtion which Gartner [2] defines as *the use of digital technologies to change a business model and provide new revenue and value-producing opportunities; it is the process of moving to a digital business* frequently confused with digitization which is defined as *the conversion of text, pictures, or sound into a digital form that can be processed by a computer* [3]. To minimize any risk of misinterpretation this work utilizes the term ALIZA to clearly specify the focus on DigitALIZAtion. The implementation during this study is specific to the Manufacturing sector but the ALIZA canvas, process and tools have been designed to be generic in nature and may find applications in many other domains.

The Free Model (Google, Facebook), is an excellent example of a disruptive digital business model. It disrupts the market with an *"if-you are-not-paying-for-the-product-you-are-the-product"* ethos. To achieve this business model, vendors radically migrated their primary function from *"providing excellent products"* to *"harvesting data from excellent free products"*. New digital business models in the manufacturing sector may need to be just as radical. But how will such radical changes be communicated to the complete manufacturing organization? A canvas is required on which to paint the big picture which conveys the complete transition to Digitalization; the requirement for The ALIZA Canvas emerges.

The ALIZA Canvas in isolation is not sufficient for the management of this transition. It explains the *Why, What & How* but it has no sense of *When* or *Who*. This work suggests that the *When* and *Who* is best conveyed by the utilization of a roller coaster diagram called The ALIZA Process which conveys the kinetic energy that can be provided by utilizing the appropriate tools to engage the relevant stakeholders at the correct time in the process. With the canvas and process defined the final step is to design The ALIZA Tools which will be used to benchmark and control the processes. The primary function of these tools is to benchmark and stimulate analysis to discover potential for improvements. With that in mind they must provide a scale or continuum to enable the participants to determine where they are now (Current Score) and where they want to be (Target Score). This study *"Defines"* the key Attributes for the stakeholders so that they can actively *Measure, Analyze, Improve and Control* them; providing the stakeholders with a Six Sigma process.

The methodology utilized to design the above stated ALIZA canvas, process and tools and their application to I4 Manufacturing sector is outlined detail over the following sections.

2.2. The design of the ALIZA canvas

The Eisenhower Decision Principle (EDP) has been popularized by Covey [4] in the creation of a *four quadrant Importance / Urgency Matrix* which is commonly utilized for time management. Leveraging Covey's four-quadrant approach to create *The ALIZA Canvas,* and supplementing it with a novel presentation procedure, it becomes feasible to efficiently bridge the communication gap between the primary and secondary stakeholders (Business and Technical Stakeholders in the manufacturing domain) as outlined in **Figure 1**.

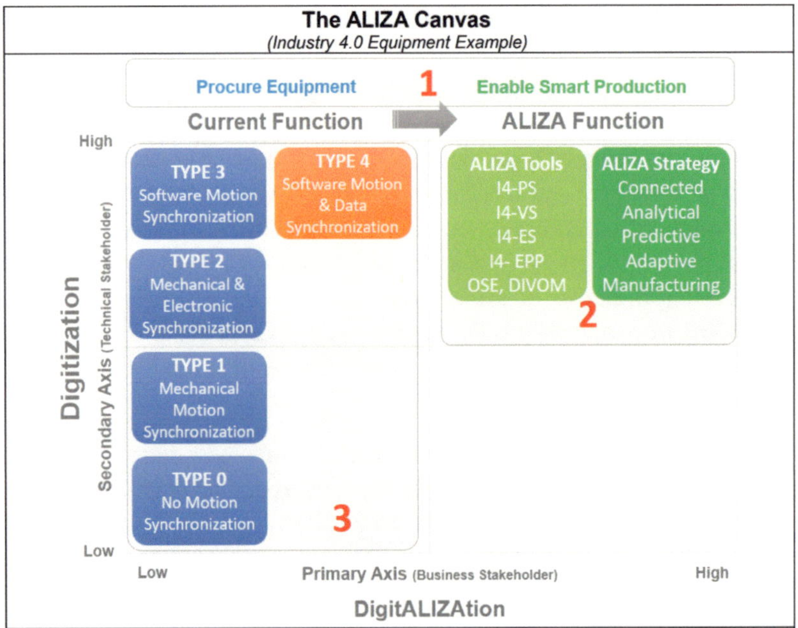

Presentation Procedure for The ALIZA Canvas

Start with the blank canvas and complete the following steps to populate the canvas:

1. **Define both the Current Function and the** ALIZA **Function**: Clearly state in business terms what the Current Function is and what the ALIZA Function will be. This is not a trivial exercise. It is extremely important that this step is owned by the Business Stakeholders, they cannot be allowed to delegate it to Technical Stakeholders. If there is too much technical influence, the definition of The ALIZA function will be technology as opposed to business centric
2. **Specify the** ALIZA **Strategy and select the** ALIZA **Tools**: Clearly state in business terms how the ALIZA Strategy will deliver the ALIZA Function. In this example *Connected, Analytical, Predictive and Adaptive Manufacturing Processes will enable Smart Production*
3. **Specify the Technology Types**: Informed by the ALIZA Strategy and provided with the ALIZA Tools the Technical Stakeholders can specify the types of the Technologies and position them appropriately so that it is immediately obvious which Technologies are enablers for achieving the ALIZA Function. In this example Type 4 equipment is the real enabling technology.

Figure 1. The ALIZA canvas and procedure for population.

The compact format of The ALIZA Canvas empowers the Business Stakeholder to rapidly communicate *The ALIZA Function*, which must be achieved by the business, how *The ALIZA Strategy*, will deliver it and *The ALIZA Tools* which will be utilized to select the appropriate technologies. The Technical Stakeholders categorize the type of digitization technologies and select the most appropriate methods of achieving *The ALIZA Function.* It paints a clear picture for all the stakeholders, without focusing on the fact that they talk extremely different languages. The adage *a picture paints a thousand words* applies.

2.3. The design of the ALIZA process

When *The ALIZA Canvas* is supplemented by *The ALIZA Process* (**Figure 2**) a true sense of *Who* must use *Which* tools at the different stages of the process becomes apparent. The roller coaster format is ideal for conveying who provides the energy is to enable the process (i.e. the Business, not the Technical Stakeholders) while the loops portray the critical requirement for collaboration between the various stakeholders at specific stages in the process.

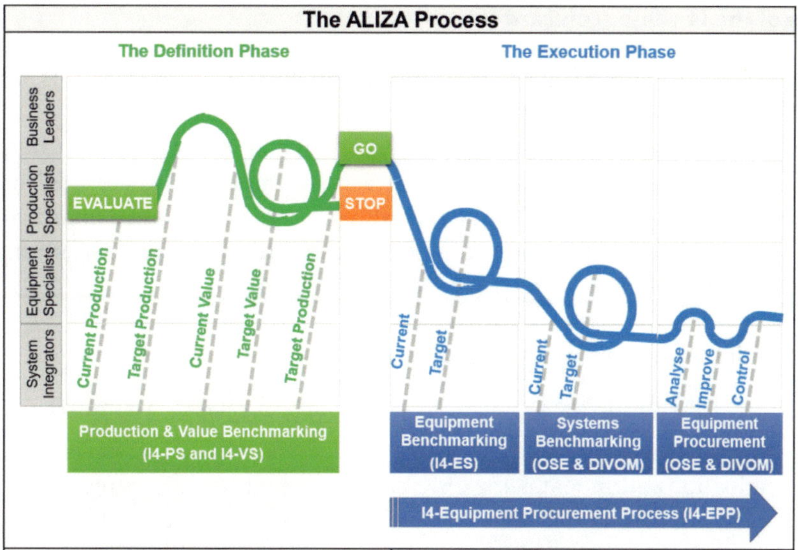

The Definition Phase	The Execution Phase
1. The starting point should be driven by Production Specialists who **EVALUATE** what is possible with Production (using the I4-Production Scorecard (**I4-PS**)). 2. Then they must work with the Business Leaders to determine what value can be delivered to the Business (using the I4-Value Scorecard (**I4-VS**)) 3. Finally they should adjust their vision for I4 Production based on the value proposition before finalizing the strategy and communicating it to the other stakeholders. 4. When Business Leaders are confident that sufficient value can be derived from The ALIZA Strategy, they should **GO** to the execution phase, otherwise they should just **STOP** and not proceed further.	1. The execution phase requires a comprehensive Industry 4.0 Equipment Procurement Process (**I4-EPP**) 2. In the I4-EPP the required equipment is benchmarked (using the I4 Equipment Scorecard(**I4-ES**)) 3. The Overall Systems Effectiveness (OSE) Calculator is then utilized to ensure that the **D**esign of the I4-EPP has the appropriate level of **I**ntegration, to ensure that the specified needs of the **V**alidation, **O**perations and **M**aintenance are met (**DIVOM**). 4. The OSE Calculator and DIVOM benchmarking tools are used throught the Equipment Procurement phases in order to ensure the the equipment remains aligned to the requirements.

Figure 2. The ALIZA process explained.

2.4. Design of the I4 production scorecard (I4-PS)

The VDMA represents more than 3200 mostly medium-sized Companies in the capital goods industry in Germany, making it the largest Industry Association in Europe. The VDMA have produced guiding principles for the implementation of Industry 4.0 in small and medium sized businesses by utilizing *Toolbox Industry 4.0 for Production and Product* [5]. These toolboxes have valuable graphical content, but they also have an underlying process which makes them even more valuable. The Production toolbox utilizes a matrix of six functions arranged in rows and progressively more favorable status arranged in five columns. By assigning a numeric score to each column it becomes possible to create a numeric scorecard namely I4-PS Scorecard for Production. This approach also facilitates the utilization of numeric *Current* and *Target* scores to facilitate the quantification and management of improvement. A detailed analysis of the content of the I4-PS Scorecard reveals that the icons from left to right can be classified as *not connected, partially connected, fully connected, analyze & predict* and *adaptive* across each of the six functions contained in the rows. These classifications in conjunction with the scores assist the rapid formulation and communication of *The ALIZA Strategy* in a concise and accurate format.

2.5. Design of the I4 value scorecard (I4-VS)

Even though this work identified the requirement for an I4-VS scorecard this work did not focus on its creation. The definition of value for a Business is a specialized domain well outside the scope of this research. Generic tools such as the McKinsey Digital Compass which maps the Industry 4.0 levers to the key value drivers [6] are freely available. Each company's definition of value will almost certainly be quite different and confidential thus it very unlikely that it will be possible to provide a generic I4-VS scorecard. Based on this assumption, each organization should define its own specific I4-VS scorecard.

2.6. Design of the I4 equipment scorecard (I4-ES)

On first impressions it appeared that the VDMA's *Toolbox Industry 4.0 for Product* [5] is not relevant to this study because this study is not focused on the product which is being manufactured. The production equipment is the Original Equipment Manufacturer's (OEM) Product, so it is relevant. By assigning the same numeric score to each column it becomes possible to create a second numeric scorecard namely I4-ES Scorecard for Equipment. This approach also facilitates the utilization of *Current*, and *Target* scores to enable the quantification and management of improvement. Unfortunately, the common denominators of not *connected, partially connected, fully connected, analyze & predict* and *adaptive* which are on the I4-PS do not map directly the I4-ES. Even though it is tempting to change some of the content of the VDMA Toolboxes this has been resisted to ensure that adherence to the VDMA's best working practice is always retained. The minor embellishments proposed in these sections only assist with the usability of the tool but do not compromise the integrity of the content. It is important to note that the I4-PS and I4-ES are owned by the Business and Technical Stakeholder's respectively. This provides a very similar format of scorecard which will undoubtedly assist in bridging the gap between the two domains.

2.7. Definition of the equipment types

An in-depth review of leading assembly and packaging equipment Original Equipment Manufacturers (OEMs) revealed a common denominator; *They all transport and perform actions on the product* [7]. A review of the logic and motion technology providers also revealed a common denominator; *they all recommend the utilization of servo motors with decentralized drives synchronized via motion control networks as opposed to mechanical synchronization* [7]. But not all equipment types require motion synchronization between the transport system and stations organized around the transport system. It is perfectly valid to have equipment which operates in an asynchronous or semi-synchronous fashion. At the other extreme, there is a growing requirement to synchronize process data between the stations and transport system. Using the technologies which synchronize the transport and stations as a separator this work defined and published a novel classification method for the different Equipment Types whereby *0—No motion synchronization, 1—Mechanical motion synchronization, 2—Mechanical & electronic motion synchronization, 3—Software based motion synchronization and 4—Software based motion & data synchronization* [7].

2.8. Definition of the equipment procurement processes (EPP)

With an Industry 3.0 EPP, the mechanical discipline typically drives the process. The Information Technology (IT) and Information Systems (IS) infrastructure are not installed, simulated or tested at the OEM's premises, thus it is not possible to test many of the critical functions at Functional Acceptance Test (FAT). This results in an undesirable situation whereby many equipment defects only become apparent after the equipment is in production. Such defects are extremely expensive, and sometimes impossible, to rectify when the equipment is in production, where limited OEM support is available. These defects undoubtedly have a significant negative impact on OEE and regulatory compliance during production. This EPP is undesirable for any equipment type but it totally unsuitable for Type 3 and Type 4 and thus cannot be utilized for Industry 4.0 Equipment.

Industry 4.0 has enabled significant advances in Industrial IT, Internet based collaborative technologies and cloud computing. These advances have all but eliminated the historical infrastructural constraints which I3-EPPs were exposed to, because it is now technically possible to simulate virtually any IT or IS, in the form of an I4 Infrastructure, at the OEM's site. The provision of an Industry 4.0 infrastructure for the FAT does not, in isolation, address all the issues which have been identified during this research. The unacceptable level of software defects which exist in *custom software* [8, 9] justifies the utilization of an *Integrated Software Quality Tool* [10], which focuses on requirement risk, test and defect management during the construction of the equipment. By adding Information Technology Infrastructure Library (ITIL®) into the scope of Integrated Software Quality, a Service Desk can be provided which facilitates the efficient provision of incident, problem and change processes to manage Data-Information-Knowledge-Wisdom (DIKW) [12]. The inclusion of these tools in a novel fashion enables the creation of an I4-EPP, as outlined in **Figure 3**. This is significantly more holistic than the I3-EPP and enables the creation of a collaborative supply network "*In collaborative supply networks, OEMs will be able to offer value-added services (e.g. maintenance, upgrade) or even sell their 'products as a service'. Remote service management helps to improve*

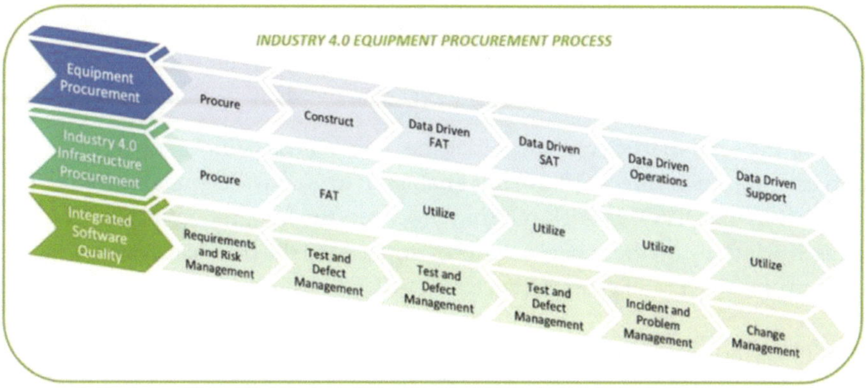

Figure 3. The Industry 4.0 equipment procurement process (I4-EPP).

equipment uptime, reduce costs for servicing (e.g. travel costs), increase service efficiency (e.g. first-visit-fix-rates) and accelerate innovation processes (e.g. remote update of device software)" [11].

2.9. Design of DIVOM and OSE

The increased complexity of Equipment Types 3 and 4 require the Concurrent Engineering approach inherent in the System Engineering process [12] to minimize costly mistakes late in the EPP. In the same way that products must be *Designed for Manufacturing and Assembly* (DFMA) [13] I4-EPPs must be designed and not simply left to chance if they are to succeed.

By applying Quality Function Deployment (QFD) [14], to the I4-EPP this study established both the high level and detailed functional requirements required to deliver the customer requirements of data driven acceptance test, excellent regulatory compliance, high OEE in production, and a fast OEE ramp. The House of Quality (HoQ) structure at the core of QFD is widely accepted by product designers, but it is a complex format and requires considerable effort to achieve an acceptable level of familiarity. This study did not consider the HoQ a suitable method of communicating the status of requirements with the various disciplines of an I4-EPP team. This study created an optimized format by taking the outputs of the QFD process and grouping the design requirements by department as follows:

> *The Project Manager is responsible for the* **D***esign of the I4-EPP to ensure that the engineers utilize the appropriate level of* **I***ntegration to meet the specified needs of the* **V***alidation,* **O***perations and* **M***aintenance customers. There we have it;* ***DIVOM*** *for I4-EPP like we have DFMA for products.*

DIVOM provides each Department with the ability to focus on the relevant design requirements of the equipment from their departmental perspective by enabling the definition of clear boundaries of responsibility but simultaneously maintaining a concurrent, cross Department focus on the complete design requirements and I4-EPP objectives.

The design of the I4-PS and I4-ES scorecards are appealing to users because they do not; (1) have too many choices, (2) require too much thought or (3) suffer from lack of clarity. These three factors are extremely important because they all increase *cognitive load* [15] in short term memory which is only capable of holding *Seven, Plus or Minus Two Objects* [16]. The DIVOM process, which is based upon thousands of requirements, had to be broken down into several stages to achieve similar levels of *simplicity* but also must be capable of rolling up to provide an overall Key Performance Indicator (KPI) which can be easily explained and understood.

An analysis of the OEE metric concludes that the *"understandability"* of the metrics [17, 18] as opposed to its *numerical accuracy* [19, 20] has enabled it to gain widespread acceptance. The key technical metric in the DIVOM benchmarking process is the Integration metric. The integration metric focuses primarily on the cyber systems as opposed to the physical equipment. This led to the suggestion of the Overall Systems Effectiveness (OSE) metric which was defined as follows:

$$\text{OSE} = \text{Design}^{*}\,\text{Average (Integration, Validation, Operation, Maintenance)}. \quad (1)$$

This formula for the OSE proposes that the Design of the I4-EPP is the enabler for achieving the highest level of OSE and as such must have the single biggest influence. Each of the other metrics are of equal importance but will have less overall impact than the Design metric. This research is not suggesting that such a simplistic formula is capable of accurately representing every situation and it will undoubtedly require future refinement. But in its current format it leverages the lessons learned from OEE and is sufficient to provide a quantifiable benchmark metric which is fit for purpose.

A hierarchical structure and an executable application (The OSE Calculator) were developed to support the DIVOM process. The OSE Rating is at the top of the hierarchy. It is composed of five metrics (D, I, V, O and M) each with three components consisting of 10 Attributes. Attributes are composed of a variable number of requirements which are omitted from The OSE Calculator application to minimize the cognitive load on participants. It is important to note that omitting the requirements significantly increase the dependence on the facilitator.

The 10-Attribute scale is organized in order of achievement with 00 being worst to 10 being best in class. This approach has been utilized to expedite user comprehension of the measurement process. The clarity of the Attributes is further augmented by adoption of the standard color coding convention of green = low risk, orange = medium risk, red = high risk and displaying the rating graphically (see **Figure 4**).

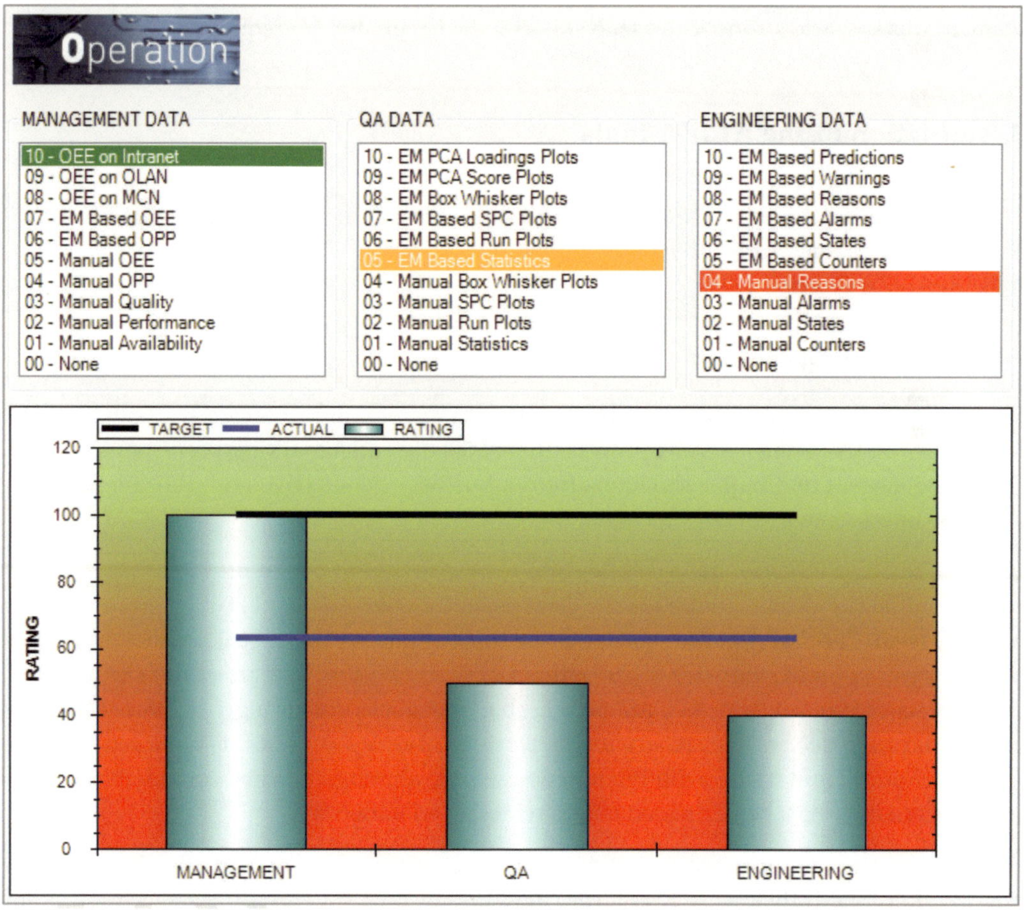

Figure 4. Design of the OSE calculator.

A three-step process was utilized with The OSE Calculator to score the metrics for the calculation of the OSE Rating; (1) Specify the ***Validation, Operation*** and ***Maintenance*** customer requirements, (2) determine the appropriate level of ***Integration*** (3) ensure that the ***Design*** of the EPP is correct. The metrics are evaluated by examining each component's Attributes in turn (from 00 to 10) to determine which Attributes will be achieved (see **Figure 4**).

Designing The OSE Rating and The Calculator in this fashion significantly increased the potential of conveying a large amount of specialized requirements to a general audience in an extremely short time period and providing five key metrics and an overall KPI to enable a Six Sigma approach to the EPP.

Even though the 5, 3, 10 formats of Metric, Component and Attribute creates a uniformity to reduce the cognitive load it introduces a constraint which although not immediately apparent may be problematic in some situations. The constraint is that the Attributes in The OSE Calculator may not be applicable in every situation, thus there is a requirement for the facilitator to state this as they navigate through the process. Another observation was that the participants frequently wanted to "score high". When these two items are combined they frequently attempt to utilize various justifications to claim that critical Attributes are not applicable to them. In this scenario strong leadership skills by the Facilitator are required.

3. Validation of the ALIZA tools

The experiments outlined in the following sections were conducted to validate that *The ALIZA Tools* have a suitable *"Form"* to reliably deliver the required *"Function"* and seamlessly *"Fit"* into *The ALIZA Process*.

3.1. The scorecards method

Regarding *Fit*, it is no stretch of the imagination to state that the scorecard methodology *"fits like a glove"* in the overall ALIZA process. It enables the benchmarking processes to rapidly define the current and target situations from a business perspective and utilize a relatively seamless interface to The OSE Calculator.

But the *Form* and *Function* are a totally different matter. Most of the Business Users, after a cursory initial inspection, jumped to the conclusion that the I4-PS and I4-ES scorecards are in an ideal form. The fact that the scorecards clearly outline the key metrics, which have been defined by a reputable industry organization (VDMA), the graphics are easy to understand, they have a continuum and they are measurable appealed very strongly to them. One even went so far as to declare *"Great, now we can manage Industry 4.0"*. But this work has delved a lot deeper and found issues with the function which, although not insurmountable, are not insignificant and must be addressed before widespread adoption.

An experiment was designed to validate the accuracy of these scorecards. The objective was to determine if they were *Repeatable* (the same inspector getting the same result when evaluating the same item more than once) and *Reproducible* (inspectors getting the same result when evaluating the same item) as gauges [21]. The first stage of validating the scorecards was conducted at the end of the first semester on the MEng in Mechatronics, University of Limerick, 2017. Eight students worked as a group and utilized the I4-PS and I4-ES to rate two pieces of equipment. The second stage of validating the scorecards was conducted at the end of the second semester. Four random students, who were members of the original team, were requested to utilize the I4-PS and I4-ES again to rate the same two pieces of equipment. The results were analyzed, and significant variation was observed. On the five-point scale of the scorecards the Lower Control Limit (LCL) lay close to 0 across all metrics and equipment while the Upper Control Limit (UCL) ranged between 3 and 5. Several factors such as group dynamic versus individual score, new knowledge attained, knowledge forgotten or simply confusion may have influenced these outcomes. Regardless of the root cause of the variability these results do highlight the fact that gauges which appeal to our desire to not increase our cognitive load [15] and are easy to memorize [16] in no way guarantee that they are accurate.

But all is not lost. A detailed review with the students revealed that they had significantly different interpretations of the iconography, the words simply were not descriptive enough and open to interpretation (e.g. What does" *connected*" really mean?) while many were not *mutually exclusive*. Thus, it can be concluded that with further experiments the content of the scorecards can be optimized to minimize variability and increase the accuracy to a point whereby the scorecard methods can be generally relied upon to achieve their *Function*.

3.2. The OSE calculator and DIVOM method

The first stage of the validation of The OSE Calculator and DIVOM Method focused on four industrial EPPs from 2012 to 2016. During these EPPs the researcher performed a DIVOM assessment and facilitated OSE Optimization sessions which evaluated how useful the participants found the overall tools and process. Informal interview and data capture techniques were utilized throughout these sessions.

The case studies clearly demonstrated that DIVOM benchmarking process achieved its *Function* of delivering tangible business benefits in the form of a Data Driven FAT, increased OEE and improved regulatory compliance, but with two strict provisos; the Project Sponsor must be a Change Agent focused on Industry 4.0 (Case Study 1 and 4). If the Project Sponsor is not empowered to enact change (Case Study 3) or is a diehard I3-EPP supporter (Case Study 2), then these methods are worthless and should not be utilized. Even though general awareness of I4 should have progressed since the recommendations were published [22], this work has uncovered underlying inhibiting factors which must be addressed.

Most specialists, observed during these case studies, were unwilling to gain an understating of an Attribute which they felt was not part of their primary discipline. It appears they were intimidated by having to admit that they needed to learn about these Attributes. They were *"the teachers"* not the *"students"*. They were extremely quick to disown these Attributes and assign them to other disciplines without personally gaining any knowledge. Even though it is outside the scope of this stage of the research, this reaction presents an insurmountable barrier to transdisciplinary [23] collaboration and must be better understood if I4 is to succeed in an efficient fashion.

The second stage of the validation of OSE Calculator and DIVOM Method focused on the 2016 and 2017 MEng in Mechatronics at the University of Limerick. The objective during this stage was to determine if the *Form* of the DIVOM process was suitable. A high degree of confidence had been gained from the case studies that the form of The OSE Calculator was fit for purpose in the hands of skilled facilitator, but the question which had to be answered was if others could be trained to be confident Facilitators? This stage did not focus on measuring the absolute accuracy of the student's knowledge because of the risk of bias based on association with academic grading. Instead the students were requested to estimate their own level of understanding to determine their *"confidence"* level. This assumes that any inaccuracies could be minimized based on further training if required.

The same academic format was utilized in the 2016 and 2017 classes. The students were not given access to The OSE Calculator at the outset. They were provided with the Attributes grouped by Component and Metric in a Microsoft Excel Workbook. The 2017 students were provided with a Microsoft Word Document containing explicit requirements for each Attribute at the start of the year, while the 2016 students were not provided with the explicit requirements. In the first semester the theory behind the DIVOM Metrics, Components and Attributes were explained and the students were mentored as groups to perform a DIVOM assessment on the group EPP. In the second semester they worked individually to complete the design of their solution as part of the group EPP, while in the third semester they executed the group EPP. At the end of each semester every student was requested to estimate their % understanding of each Attribute, based on the explanation that this would help to focus future lectures where the gaps in understanding existed (to mitigate the risk of students over estimating their % understanding in the hope of obtaining a higher academic grade).

All students, despite some having quite significant Industrial experience, estimated their initial understanding at close to 0%. At the end of the first semester students with access to the explicit requirements (2017) claimed to have an average of 55% understanding while those without (2016) had only 29% understanding. By the end of the second semester this gap virtually disappeared (67% for 2016 and 68% for 2017) while at the end of the third semester the 2016 group had achieved a very high 78% (2017 not finished at time of publication).

The sample size of eleven completed workbooks is too small to draw definitive conclusions from, but they are adequate to provide early indications and direct further work. Even though the DIVOM Attributes may provide an ideal framework for an expert they are extremely intimidating *Form* for novices. This may go a long way to explaining the behavior of the specialists in the case studies. Detailed requirements which further explain the Attributes rapidly increase the user's perception of their understanding of the Attributes. They are very useful for reducing the intimidation factor which was observed during the case studies.

If the detailed requirements were provided as pre-reading to the attendees of an OSE Optimization workshop it may enable them to inform themselves prior to the workshop and minimize the intimidation factor. Because these requirements are at the lower levels of

Bloom's Taxonomy [24] they could easily be tested with tools such as Moodle quizzes and the student provided with a novice level certification prior to attendance. This has the potential to create a process which transcends discipline and the collaboration issues which they cause. In this scenario the *Trans-Attribute* collaboration can be enabled where *team members must be competent enough in their own* ***Attributes*** *and understand the language of all relevant* ***Attributes*** *that enables them to contribute to the members' quality research or learning and combine various perspectives to build up a new framework* [23].

This stage of the research clearly highlights that if true understanding of the Attributes is required a significant amount of time must be invested (second and third semester) to achieve the 70-20-10 rule model [25] for learning. The dialog between a tutor and student, involving several of the common alternatives is also required to produce significantly more understanding [26] than a simplistic exposition of the correct information [27]. But that is only to be expected on the journey from Novice to Expert [28].

Regarding *Fit*, gloves normally come in pairs and The OSE Calculator and DIVOM Method is the second glove which compliments the first; the Scorecard Methodology. As with gloves, one is of limited use and the whole (The ALIZA Process) it much greater than the sum of the parts (Scorecard Methodology) + (The OSE Calculator and DIVOM Method).

4. Design of E-Cubers

4.1. Introduction

The ALIZA Canvas, Process and Tools for I4 Manufacturing Equipment represents a significant output of this work, but an educational mechanism is required to rapidly disseminate these tools and methods to derive tangible benefit of industry. This section explores conventional academic educational structures and concludes that an additional, complimentary, structure for *inventing and implement technical solutions, for business problems, in the I4 equipment domain* is required. To that end, the E-Cubers organization has been created with the following objective:

> *E-Cubers is an educational organization consisting of a constellation of Communities of Practice (CoPs) organized around topics which are designed to facilitate collaboration and creativity for the advancement of each members individual competencies to support the achievement of I4 Equipment Engineering Excellence.*

Designing and implementing a constellation of CoPs is not trivial matter. In fact, it is fraught with difficulty, but the benefits can be enormous [29]. E-Cubers are only at the start of this exciting journey of exploring how the CoPs can be organized to be truly effective Knowledge Management Systems promoting effective and productive collaboration in the Industry 4.0 Equipment domain. It should not be assumed that CoPs in isolation can guarantee the creativity required for the invention of novel solutions in Industry 4.0. But what is creativity? How can it be nurtured? By examining the applicability of Resnick's Four Ps [30] for cultivating creativity, in the general sense, and refining it to the E-Cubers specific requirements this work has defined the E-Cubers Eight Ps for cultivating creativity.

4.2. Designing E-Cubers for knowledge management & collaboration

An essential first step in designing the E-Cubers constellation of CoPs is to determine the unit of organization for each CoP. An initial reaction may be to organize by discipline or subject and adopt one of the emerging trends such as antidisciplinary collaboration [31] (more frequently referred to as Trans-Disciplinary). Unfortunately, during the execution of the DIVOM case studies this work found that the term discipline, regardless of the pre-fix, can be counterproductive for an organization such as E-Cubers. Disciplines are defined as *a branch of knowledge, typically one studied in higher education* [3]. In an academic setting disciplines are organized by Department and progression is based upon an examination and individual qualification organized around an annual academic calendar. Significant status is allocated to qualification with very little reward for collaboration. In some instances, it can be so extreme to be virtually gladiatorial in nature. For this reason, disciplines are regarded as unsuitable for the promotion of collaboration by E-Cubers and will not be used.

Wenger [29] explains that at the core of CoP there must be a shared domain which creates a sense of accountability to a body of knowledge and therefore to the development of a Practice. It is not an abstract area of interest but consists of key issues or problems that members commonly experience. It is not merely a passing issue, which can be addressed by a temporary task force. It concerns complex and long-standing issues that require sustained learning over an extended period. It is essentially a *topic—a matter dealt with in a text, discourse, or conversation* [3] as opposed to a *subject—a branch of knowledge studied or taught in a school, college, or university* [3]. The word topic has been selected as the unit for E-Cubers CoPs as opposed to domain because it evokes a more succinct and narrow focus than domain. Also, topic does not infer the requirement for examination which is assumed with the word subject.

By limiting each E-Cubers CoP to a clearly defined topic it becomes possible to define the internal structure of a CoP around the five levels of competency, namely *0—None, 1—Novice, 2—Proficient, 3—Independent, 4—Advanced, 5—Expert* [28]. Utilizing this format, individuals can easily declare their level of competence in the topic, without any risk of intimidation or feeling inadequate. Levels *4-Advanced* and *5-Expert* status is achieved based on the production of knowledge assets [28]. Experts are ideally placed to create knowledge assets which helps Novices to *"understand the language"* in an extremely short time period. This enables E-Cubers to achieve Trans-Topic collaboration on the Topic Collaboration Spectrum (see **Figure 5**). This novel approach will enable E-Cubers to achieve an unprecedented level of knowledge creation and dissemination at the topic as opposed to discipline level.

This Trans-Topic collaboration approach has the potential to create an extremely powerful evolutionary knowledge management model. Even though it is not applicable to every Topic, the rate of creation of these knowledge assets can be drastically accelerated for many Topics by organizing CoP domains around technology provider vendor Topics and leveraging their resources. This practice has gained wide acceptance in the IT domain with large technology providers such as CISCO and Microsoft's certified courses being taught by conventional academic institutions. There is no reason why this approach cannot be replicated throughout the E-Cubers constellation of CoPs.

The E-Cubers Topic Collabortion Spectrum		
Trans	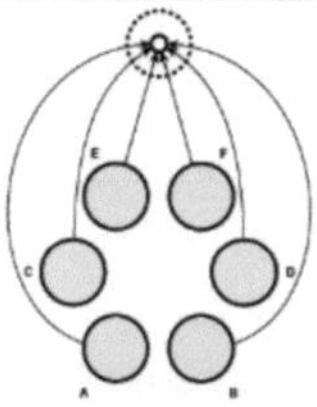	Trans-Topic collaboration is concerned at once with what is between, across and beyond all the domains with the goal of understanding the present world under an imperative of unity of knowledge. It begins in the white space between topics. There are no existing specialists, frameworks or methods and therefore the language and fabric of this unexplored area is still left to be defined and can be discussed on a level playing field.
Cross	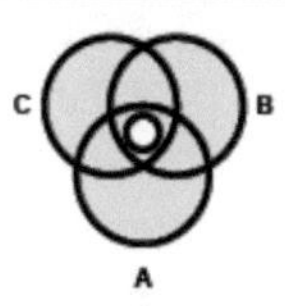	Cross-Topic collaboration is concerned with the study of a problem at the intersection of multiple topics, and with the commonalities among the topics involved. Cross-Topic knowledge is that which explains aspects of one topic in terms of another. Understanding an aspect of your topic through the lens of another provides you with a wider context for the problems you are facing and hence this will refine your approach to problem-solving
Inter	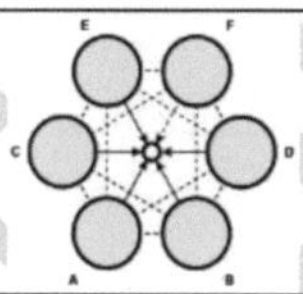	Inter-Topic collaboration is concerned with the study of a problem within multiple topics, and with the transfer of methods from one topic to another the solution integrates different topic approaches and methods.
Multi	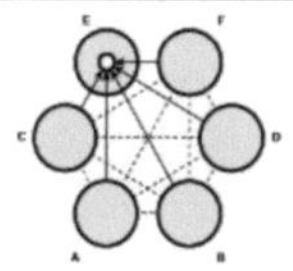	Multi-Topic collaboration is concerned with the study of a solution within one topic, with support from other topics bringing together multiple dimensions, but always in the service of the driving topic. Topic elements retain their original identity. It fosters wider knowledge, information and methods
Uni	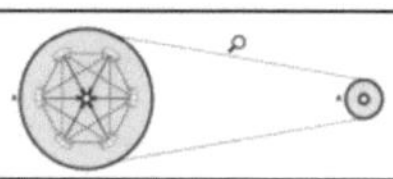	Uni-Topic collaboration is the transfer of knowledge that occurs within the confines of a defined topic e.g. Siemens programmers working on a Programmable Logic Controller (PLC) problem

Figure 5. The E-Cubers Topic Collaboration Spectrum.

E-Cubers will utilize Office 365 for Education to manage its constellation of CoPs. This means that E-Cubers can collaborate regardless of location and organization boundaries. The correct utilization of SharePoint's CoP site templates enables a Topic-centric Knowledge Management System (KMS), which is of significantly more value for E-Cubers objectives than the discipline centric Learning Management Systems (LMS) frequently utilized by Academic Institutions. This approach facilitates a transition from the existing function of *discipline centric qualifications* to the E-Cubers ALIZA function of *topic centric competence* and an E-Cubers ALIZA strategy of *World-class equipment knowledge assets, for any E-Cuber, anywhere*. The E-Cubers constellation of CoPs facilitates, for the first time, the full collaboration spectrum from Uni-Topic the whole way to Trans-Topic to solve whatever problem is at hand in the I4 Equipment domain. Now that is a truly disruptive concept!!!

4.3. Designing E-Cubers strategies for creativity

Sir Ken Robinson, in his TED Talks on creativity outlines the importance of risk taking. He outlines that *"If you are not prepared to be wrong, you'll never come up with anything original,"* and raises the criticism that. *"We are running education systems where mistakes are the worst thing you can make. We are educating people out of their creative capacities."* Essentially, we are not empowering them to invent; E-Cubers must strive to reverse this trend in the I4 Equipment domain.

At school, students are taught that they must do their own work. If they leverage the work of others it is frequently seen as cheating. This is in direct conflict with the methodologies of CoPs, the open source community, creative commons and indeed general industrial practice. E-Cubers will follow in the footsteps of MIT Media Labs who despite frequent lobbying, provides an extremely strong stance of utilizing the Creative Commons License to support the concept of sharing what has already been achieved by using the remix function in MIT's Scratch environment [30]. E-Cubers will also apply the Creative Commons License to its work and utilize the everyday creativity of *little-c*, while fostering the *mini-c* inherent in the learning process during the attainment of the professional-level expertise required to practice *Pro-c*; and the expectation of the occasional *Big-C* [32]. Thus, *Big-C* is the welcome surprise but not the primary objective of E-Cubers. E-Cubers prolific creation of valuable *little-c, mini-c* and *Pro-c* knowledge assets will further displace the conventional misconceptions that creativity is about artistic expression, belongs to just a small section of the population, comes in a flash of light and cannot be taught [30].

Resnick [30] 4Ps methodology of *cultivating creativity through **Projects**, **Passion**, **Peers** and **Play*** is a truly authoritative piece of work, but this does not guarantee complete applicability in every domain. One could argue that its main strength (its general nature) which enables it to have enormous global impact when utilized at a macro scale is a significant weakness when applied at a micro level to a single organization such as E-Cubers. E-Cubers is focused on identifying and nurturing the knowledge of *Patterners* who can support I4 Equipment as opposed to catering for the *Dramatists* and the wider needs of Society. But with a *little-c* Resnick's Four P's work can be further expanded applied to E-Cubers to great effect.

Resnick [30] outlines his surprise that the Danish language unlike English has two words to distinguish between the different types of play (*spille* and *lege*). But the same is true in Gaelic, the native Irish language. In Gaelic *imir* means to play by taking part in sport or game while *spraoi* means playing to have fun. Papert has a strong aversion to *imir* and allows participants to derive their ***Passion*** from working on ***Projects*** based around their personal interests with no boundaries. E-Cubers will instead focus on utilizing Papert's *"low floors"* to provide an easy way for novices to get started and *"high ceilings"* to allow them to work on increasingly sophisticated projects over time [33] without extending to the *"wide walls"* [30] which is too general for the specific E-Cubers objectives. E-Cubers will facilitate participants to obtain their ***Passion*** from *imir* (competing) at something which they love to *spraoi* (play) in the I4 Equipment domain. E-Cubers will practice *"ag spraoi agus ag imirt"* which translates to *"fun and play (by competing)"* in English.

I4 Equipment requires a digital twin with a representation in both the cyber and physical worlds. The cyber world is a novelty and is currently achieving enormous focus. So much so that the physical world is being somewhat neglected. If E-Cubers are to implement I4 Equipment solutions it is critical that they understand both the physical and the cyber worlds. They need to start young. The younger the better. To assist this engagement E-Cubers has created a progression sequence consisting of three distinct stages; **BUILD, PROGRAM, INVENT** which goes much further, albeit in a specific domain than purely programming centric solutions.

The **BUILD** stage utilizes LEGO® TECHNIC sets for the equipment kits. It enables participants *to "learn by playing"* by constructing the pre-designed models. It requires a strong attention to detail and organizational skills. It does not require a high level of creativity; that comes later, once the participant has grasped many of the basics. The **PROGRAM** stage utilizes LEGO® MINDSTORMS sets for the equipment kits. The participants learn about physical sensors and actuators and create programs to achieve required functions. This stage culminates in the construction the MindCub3r [34]; a piece of equipment which can solve the Rubik's Cube. The **INVENT** stage utilizes LEGO® TECHNIC sets and building instructions from PV Productions [35] for the construction of equipment based on the Great Ball Contraption (GBC). This equipment is built with standard LEGO® parts to transport LEGO® balls. It is an ideal platform for Overall Equipment Effectiveness (OEE) measurement. But it also enables two levels of invention; Equipment and Device. At the Equipment level the participant invents by creating a new design of the Equipment utilizing standard devices (LEGO® parts). At the Device level the participant invents by creating new designs of the devices utilizing freely available 3D Printing technology to rapidly manufacture them. This is true creativity at its best!!!

At the core of the OEE optimization process is the root-cause analysis of incidents to identify ***P**roblems* which must then be ***P**rioritized* to separate the vital few from the trivial many which is the essence of the Pareto Principle (even more Ps!!!). ***P**roblem* identification and ***P**rioritization* appear to have been sacrificed by Resnick for the benefit of generality, but they are key E-Cubers competencies and as such must be included. They facilitate the higher order thinking skills of analysis, synthesis and evaluation endorsed by Bloom [24] and are the starting point for the execution of the Solution based ***P**roject* which changes the design of the equipment or device to improve the OEE. This is not just *little-c* without any purpose. This is *mini-c* [32] for E-Cubers in a true *E-Cubers OEE Playground*. In the *E-Cubers OEE Playground* there is no single right or wrong answer. It is never finished. There will be a law of diminishing returns, but the OEE can virtually always be improved. Equipment such as the MindCub3r and GBCs which is capable of continuous operation are essentially *Dynamic Problem Generators* who nobody truly has the answer for. The challenge to an E-Cuber is *"Can you identify the problem, invent a solution and implement it?"* Where better for an E-Cuber to display their talents than *ag spraoi agus ag imirt* at *The E-Cubers OEE Games*? At *The E-Cubers OEE Games* the challenge is to optimize the Availability, Performance and Quality metrics of a specified piece of equipment at the three stages (**BUILD**, **PROGRAM** and **INVENT**) enabling the complete Four C Model of Creativity to be catered for [32].

4.4. Defining the E-Cubers Eight Ps

To conclude, by evaluating and expanding Resnick's Four Ps, this work defines the E-Cubers Eight Ps:

> *E-Cubers utilize the concept of an OEE **P**layground to **P**assionately build up a **P**ortfolio of **P**rojects in collaboration with their **P**eers, "ag spraoi agus ag imirt" (Gaelic for **P**laying by having fun and competing). They demonstrate their technical ability by leveraging existing knowledge assets and creating novel solutions to the **P**roblems which they have **P**rioritized as they learn the art of their craft.*

These ***Portfolios*** enable the E-Cuber to demonstrate their capabilities to both potential employers and Academia. Who knows they may eventually be regarded as a suitable assessment method by Academia? What a different world that would be!!!

5. Applying E-Cubers to the Irish education system

The Irish education system is made up of primary, secondary and third-level education. If E-Cubers is to successfully cater for the requirements of Industry 4.0 over the long term it will need to have solutions which appeal to all three of these levels. This is a significant challenge. The promotion of a culture of collaboration and creativity across all three levels of an education system to support an emerging career has simply not been done before. The following sections outline what was discovered by applying E-Cubers at each education level.

5.1. Primary level

Several E-Cubers **BUILD** workshops were held which engaged more than 100 students between 2014 and 2018. The focus at this level was to determine if the concepts of *Equipment* with *Functions* which can be *Tested* could be imparted with" *ag spraoi agus ag imirt"* as a precursor to understanding OEE. All the workshops utilized LEGO® TECHNIC pull-back racers. They cost approximately €25, have less than 150 pieces. They take less than 30 minutes for an experienced builder to build but a novice can build them in less than 60 minutes.

A very small minority of students were clearly *Dramatists* as opposed to *Patterners* and were totally averse to following the detailed instructions. But most of the students enjoyed the build process even though less than 10% of the boys and less than 2% of the girls had played with LEGO® TECHNIC before. Without exception all the students could understand the fundamental concepts of equipment, functions and testing. But their desire for *imirt* was extremely strong. They wanted to create a competition and that is exactly what we did. We created an *E-Cubers OEE Game* based on the pull-back racers which were conducted over 300 seconds. Each team of two (who had completed the BUILD process) could test the OEE of their racer on a 2-meter-long race track with a defined start line and a defined *"landing zone"*. The three standard components of availability, performance and quality were utilized to calculate OEE and the formula was explained to the students. The performance was calculated based on the number of cycles (attempts to land in the zone), the quality was calculated based on the number of good cycles (attempts which landed in the zone) and the availability was calculated based on how many seconds during the 300 second OEE Game that the equipment was working correctly. Even though there were no prizes on offer the *imirt* really did bring ***Passion*** to the ***Project*** as they collaborated with their ***Peers*** to get the highest OEE.

Even though the E-Cubers **BUILD** workshops were extremely successful at achieving the stated objective, several lessons which were learned which should be shared. Do not hold *"free events"* which are open to the public. There will be a significant variation in age groups and capabilities which make them virtually impossible to control. Where possible utilize the teacher to organize the workshop in their own school because they are a known authority figure and well respected by the students. Be very organized and ensure that all the sets have all the components. Get the students to do a stock-take before they start; this eliminates the *"a part is missing"* scenario. But most of all do not forget that you are *"ag spraoi agus ag imirt"* and enjoy the ***Passion;*** it truly is contagious.

5.2. Secondary level

Secondary Level (or post-primary) education consists of a three-year Junior Cycle (lower secondary), followed by a three-year Senior Cycle (upper secondary), if they take the optional Transition Year (TY). The TY is a critical decision point for young students as they decide on which potential career path to take. E-Cubers designed and implemented a week-long TY-PBWS (Project Based Work Simulation) program with the objective of introducing students to the Equipment Procurement Process (EPP) and outlining the different engineering roles within the equipment engineering team structure.

The TY-PBWS was provided to Coola (mixed-gender), Summerhill College (all-boys) and The Mercy College (all-girls) schools in 2014 and 2015 thus any gender imbalance should be negligible. During the TY-PBWS the students were organized in teams and assigned engineering roles. They had to **BUILD** complex equipment based on the LEGO® TECHNIC 8110 Unimog 400 or 42,030 VOLVO L350F Wheel Loader or the LEGO® MINDSTORMS based MindCub3r which solves the Rubik's Cube. They had to design and successfully execute a Factory Acceptance Test (FAT) and proceed to optimize the design of the equipment. All the teams clearly demonstrated that they could perform root cause analysis, in response to identified incidents. They demonstrated an ability to identify problems and suggest or implement suitable changes. Thus, it can be inferred that, with careful preparation and practical equipment examples, TY students can understand the key principles of both OEE and ITIL® in the equipment domain. 100% of the TY students stated that by taking part in the PBWS they obtained a better understanding of the various roles engineers play in the EPP and how EPPs are executed, while 95% claimed that they got a lot of satisfaction from getting the equipment working, which concurs with the *Constructionism Learning Theory* [33].

5.3. Third level BEng (honors) in mechatronics

The BEng (Honors) is at the cornerstone of the Third Level engineering education system. It typically takes 4 years to complete and results in a Level 8 award in the National Framework of Qualifications (NFQ). Following on from the success of the TY-PBWS, E-Cubers designed a 3-week-long BEng-PBWS and conducted at the IT Sligo, in July 2014. The BEng PBWS was focused on both the **BUILD** and **PROGRAM** of the MindCub3r equipment, but from a different perspective; to further analyze the problems affecting the OEE with the objective of developing and implementing solutions (**INVENT**).

During the BEng-PBWS, each student was made responsible for the software modules which fell within their individual scope. This resulted in clearly defined deliverables being achieved with the students demonstrating an ability to analyze, evaluate and create [24] solutions.

The complete team were responsible for the required integration testing which was scheduled to be completed in the third week. This enabled the mentor to accurately simulate the work environment. The team members were provided with time limited targets which they had to achieve, while being held responsible for a clearly defined role. All the students stated that they found the BEng-PBWS workshop format much more interesting, and that they learned significantly more than conventional academic lectures or tutorial-based environments. These findings strongly concur with Muller [26] and the 70-20-10 rule model for learning [36].

The BEng students went significantly further than just completing the BEng-PBWS. Under their own initiative they developed and proposed a complimentary pedagogy to augment the existing academic model. In this new model they recommended industrial engagement should occur in September and November during the first year to outline to student's what industry and employers really require. They also recommended workshops of three-week duration in June, July and August of first second and third year respectively. They advised that the third-year workshop should be competitive (*imirt*), only offered to high achievers and they should be paid a stipend during this workshop as they are mentored to identify and develop the concept of a truly significant final year project. They also requested that Industry provides them with periodic supervision during the execution of their final year project. This novel model which was proposed *"by BEng students for BEng students"* delivers more than 250 direct contact hours to students at key milestones in their development, which would significantly enhance their competencies and development.

This clearly demonstrates that skilled and motivated students at both TY and BEng levels are capable of both finding technical solutions and highlight business problems (albeit the business of providing a BEng qualification) which is extremely encouraging. It is important to note that all the participants were volunteers, they were not voluntold. This PBWS workshop was something they wanted to do. The results could have been significantly different if the PBWS method was applied to the full population or there was forced attendance.

5.4. Third level MEng in mechatronics

The MEng qualification is close to the pinnacle of the Third Level engineering education system. It typically takes two academic years to complete and results in a Level 9 award in the National Framework of Qualifications (NFQ). Following on from the success of the BEng-PBWS E-Cubers designed a three-semester-long MEng-PBWS and delivers it on the MEng in Mechatronics course at the University of Limerick since 2016. The MEng-PBWS is focused on *inventing and implement technical solutions, for business problems, in the I4 Equipment domain*. This means that the MEng applies The ALIZA Process and Tools defined in this study to industrial equipment as opposed to the LEGO® equipment (but the engineering principles are the same).

The subjects in the MEng-PBWS have been organized across three semesters. In the first semester the focus is on updating the students with the relevant standards for equipment such as ISA-S88, S95, S99, The ALIZA Canvas, The ALIZA Process and The ALIZA Tools (namely I4-PS, I4-VS, I4-EPP, The OSE Calculator and DIVOM) developed during this study. This semester provides the students with the basic vocabulary for ALIZA. In the second semester the focus is on mentoring the students to create a Solution Overview to form the basis of a project which they must deliver in the third semester. The objective is to assist the students to migrate from BEng to MEng level whereby they could both discover problems and define solutions. The results were quite surprising.

A significant number of students had virtually no interest in participating in the divergent stages of the *Double Diamond Design Process* [37]. They did not want to actively engage in the Discovery or Definition Stages. They simply wanted to be provided with the definition of the ***Problem*** and then they would participate, ideally in isolation, in the Development and Delivery stages. They were extremely reluctant to engage with their ***Peers*** in the form group projects, ***Passion*** for the topic (the same was true with other subjects) was virtually non-existent, so the creation of a ***Portfolio*** was a non-starter. They were extremely ***Passionate*** about obtaining the MEng qualification, the career prospects were (incorrectly in my opinion) assumed, and the topic was virtually irrelevant to them. What a shame!!!

But there were several students who were much more open to ***Problem*** definition and ***Prioritization*** activities. These students actively ***Played*** with the technologies. They closely collaborated with their ***Peers*** and delivered world-class solutions and collaborative ***Projects***, they leveraged the E-Cubers Office 365 environment and are well on the way to creating very impressive ***Portfolios***. They were ***Passionate*** about the topic(s) not the qualification.

Various factors such as personality types, entry grades, age profile and industrial experience were evaluated. The only metric which separated these two groups was industrial experience. But that simply appears to be an indicator of where the individual derives their motivation from as opposed to the explicit value of the experience. It appears that a BEng student who continues directly to a taught MEng, without any industrial experience, may be solely motivated by the extrinsic reward of the MEng qualification. This appears to concur with Pink [38] who outlines that *"Conventional thinking is that the higher the reward the higher the performance but once you get above rudimentary cognitive skill it is the other way around."* While the student with the industrial experience would appear to be engaged with the topic(s) and is truly motivated by *"Autonomy, Mastery and Purpose"* [38]. Basically, students with the industrial experience appear to be *volunteering to learn the topic* while the other students are *voluntold to learn the topic (to obtain the qualification)*. It all appears to center around where individuals derive their ***Passion*** from.

Does the solution lie in catering for each passion with different academic delivery models? It appears that the volunteers are not content unless they get to work on the topic(s). This is ideally suited to the apprenticeship delivery model whereby 30% of the training occurs in an academic environment but 70% of the competency is demonstrated in the working environment; they are the true practitioners of the topic(s). The voluntolds on the other hand want and must be catered for with the conventional but more collaborative academic model. While the true researchers can be catered for with the standard research model centered around very specific or a small number of topics. This enables the definition of three different roles based on the delivery model; namely *Equipment Systems Engineer, Equipment Systems Designer and Equipment Systems Researcher*. By letting the individual pick the role which matches their ***Passion*** and they can all significantly add value, in very different ways, to the E-Cuber Topic(s).

5.5. The E-Cubers library: a technical solution to a business problem

A surprising revelation during the execution of PBWS, which spanned the full educational system and crossed both genders was that only a very small number had access to the LEGO® TECHNIC and LEGO® MINDSTORMS kits. Following the PBWS more than 90% of students claimed that they would utilize LEGO® TECHNIC or LEGO® MINDSTORMS but less than 5% had access to them. Those that had access to these products personally owned them, which is quite a privileged situation because they typically cost between €100 and €300 for the larger sets. In contrast quick review of the E-Cubers kits which have been purchased for this research reveals a utilization of less than 5% (they are just sitting on a shelf not utilized for 95% of the time). This raised the question *"How can E-Cubers make their kits more generally available?"*

People are generally reluctant to lend LEGO® sets because parts will get lost or broken and then the set is useless. Well that was the case in the past, but now. With the advent of digital technologies there are catalogs of the parts in every set available [39], there are methods of utilizing 3D Printing techniques to replace some parts [40] and there are websites where individual replacement parts can be ordered [39]. All that is needed is the stock-take application and a method of providing the E-Cubers Kits to the public. E-Cubers have sponsored the development of the stock-take application and in 2018 E-Cubers in conjunction with Sligo County Library will be launching *The E-Cubers Library* to make E-Cubers Kits available to library members in Sligo. All the library members must do is stock-take the E-Cubers Kit before they check it out of the library and stock-take it again before the check it back into the library. There will be 3D Printers available to print replacement parts and a facility to order some replacement parts if required. When *The E-Cubers Library* has been fully implemented and proven in Sligo E-Cubers will make it available for all the 32 counties of Ireland so that the E-Cubers Kits can be available to any library member in Ireland.

Would not that be a significant achievement and a novel technical solution to address an urgent business problem (the emerging skills shortage for Industry 4.0 Equipment). It really is an ideal practical example of DigitALIZAtion and E-Cubers practicing what they preach. It can provide the resources for *E-Cubers OEE Playgrounds* where young people can foster their creativity *"ag spraoi agus ag imirt"* as they prepare for the *E-Cubers OEE Games.*

6. Conclusion

The presumption by many is that the challenges of Industry 4.0 will be predominantly technical in nature, but unfortunately that is not the case. The implementation of Industry 4.0 requires us as practitioners to fundamentally change the way in which we work. We must transcend our disciplines and collaborate at a topic level. But the changes go much deeper. We must also change how we educate our young people by creating supplementary environments which truly foster their creativity as well as their ability to collaborate. Only when we have made these changes will we truly unleash the powers of processes and tools such as *The ALIZA Canvas*, *The ALIZA Process* and *The ALIZA Tools* in the Industry 4.0 Manufacturing domain and in any other domain which we wish to apply them.

http://dx.doi.org/10.5772/intechopen.80347

Author details

Shane Loughlin

Address all correspondence to: shane.loughlin@ecubers.ie

E-Cubers, IDA Industrial Park, Collooney, Co. Sligo, Ireland

References

[1] Carroll D. Factory of the Future will See Merging of Virtual and Real Worlds [Online]. 2014. Available from: http://www.engineersjournal.ie/factory-of-the-future-will-see-merging-of-virtual-and-real-worlds/ [Accessed: Aug 29, 2015]

[2] Gartner. Gartner IT Glossary [Online]. 2018. Available from: https://www.gartner.com/it-glossary/digitalization/ [Accessed: May 8, 2018]

[3] Oxford English Dictionary. Oxford English Dictionary [Online]. 2018. Available from: https://en.oxforddictionaries.com/ [Accessed: May 8, 2018]

[4] Covey SR. The 7 Habits of Highly Effective People. Free Press; 1989

[5] VDMA & Partners, editors. Guideline Industrie 4.0—Guiding Principles for the Implementation of Industrie 4.0 in Small and Medium Sized Businesses. VDMA Verlag; 2016

[6] McKinsey & Company. Manufacturing's Next Act [Online]. 2015. Available from: https://www.mckinsey.com/business-functions/operations/our-insights/manufacturings-next-act [Accessed: May 8, 2018]

[7] Loughlin S, McGrory J. A method of classifying industrial assembly and packaging equipment. In: 2013 IEEE 11th International Conference on Industrial Informatics (INDIN 2013). Bochum: IEEE; 2013

[8] CDRH, FDA. General Principles of Software Validation; Final Guidance for Industry and FDA Staff [Online]. 2002. Available from: http://www.fda.gov/RegulatoryInformation/Guidances/ucm085281.htm [Acessed: Aug 29, 2015]

[9] Martin KC, Perez A. GAMP 5 quality risk management approach. Pharmaceutical Engineering. 2008;**28**(May/June):1-7. Available from: http://www.gmp1st.com/galleries/08mj/08mj-martin.pdf

[10] Gartner. Magic Quadrant for Integrated Software Quality Suites [Online]. 2014. Available from: http://www.gartner.com/technology/reprints.do?id=1-20U6CHY&ct=140904&st=sb [Accessed: Sep 2, 2015]

[11] European Commission. Factories of the Future Multi-Annual Roadmap for The Contractual PPP under Horizon 2020 [Online]. 2013. Available from: http://effra.eu/attachments/article/129/Factories%20of%20the%20Future%202020%20Roadmap.pdf [Accessed: Sep 1, 2015]

[12] Martin JN, Bahill AT. Systems Engineering Guidebook: A Process for Developing Systems and Products. Systems Engineering Series; 1996

[13] Boothroyd Dewhurst Inc. DFMA Design for Manufacture and Assembly [Online]. 2010. Available from: http://www.dfma.com/resources/dfmawp.htm [Accessed: Aug 30, 2015]

[14] Hauser JR, Clausing D. The House of Quality. [Online]. 1998. Available from: https://hbr.org/1988/05/the-house-of-quality [Accessed: Aug 30, 2015]

[15] Sweller J. Cognitive load during problem solving: Effects on learning. Cognitive Science—A Multidisciplinary Journal. Apr 1988;**12**:257-285. Available from: https://pdfs.semanticscholar.org/d88c/481743db95687bf9d2861c16cd006f67a0a1.pdf

[16] Miller GA. The magical number seven, plus or minus two: Some limits on our capacity for processing information. Psychological Review. Mar 1956;**63**:81-97. Available from: http://www.sns.ias.edu/~tlusty/courses/InfoInBio/Papers/Miller1956.pdf

[17] Vorne Industries. The Fast Guide to OEE [Online]. 2002. Available from: http://www.vorne.com/pdf/fast-guide-to-oee.pdf [Accessed: Aug 30, 2015]

[18] Informance. Manufacturing Benchmark [Online]. 2009. Available from: http://www.informance.com/benchmarks/ [Accessed: Feb 14, 2009]

[19] Loughlin S, McFeeley G. Applying ISA 99 & ISA 88 Principles: An Often Overlooked Aspect of Manufacturing & Packaging Systems Integration. Atlanta: WBF; 2006:1-24

[20] Loughlin S. A holistic approach to overall equipment effectiveness (OEE). Computing and Control Engineering. 2003;**14**(6):37-42

[21] Pyzdek T. The Six Sigma Handbook—A Complete Guide for Green Belts, Black Belts and Managers at All Levels. New York: McGraw-Hill; 2003

[22] Acatech. Recommendations for Implementing the Strategic Initiative INDUSTRIE 4.0. Final report of the Industrie 4.0 Working Group. Acatech; 2013

[23] Stock P, Burton RJF. Defining terms for integrated (multi-inter-trans-disciplinary) sustainability research. Sustainability, 2011;**3**:1090-1113. Available from: https://pdfs.semanticscholar.org/a356/a7d8086b5d85e7804b7d25d421520562309d.pdf

[24] Anderson LW et al. A Taxonomy for Learning, Teaching, and Assessing: A Revision of Bloom's Taxonomy of Educational Objectives. Abridged ed. Pearson; 2000

[25] Lombardo MM, Eichinger RW. The Career Architect Development Planner. 3rd ed. Lominger Limited; 2001

[26] Muller DA. Designing effective multimedia for physics education [online] (PhD thesis). 2008. Available from: http://www.physics.usyd.edu.au/super/theses/PhD(Muller).pdf [Accessed: Aug 1, 2014]

[27] The Kahn Academy. Kahn Academy. [Online]. 2018. Available from: https://www.khanacademy.org/about [Accessed: May 8, 2018]

[28] Engineers Ireland. CPD Accredited Employer Standard Support Guide. Dublin: Engineers Ireland; 2001

[29] Wenger E, McDermott R, Synder WM. Cultivating Communities of Practice: A Guide to Managing Knowledge. Harvard Business School Press; 2002

[30] Resnick M. Lifelong Kindergarten, Cultivating Creativity through Projects, Passion, Peers and Play. London: The MIT Press; 2017

[31] Ito J. Antidisciplinary [Online]. 2014. Available from: https://joi.ito.com/weblog/2014/10/02/antidisciplinar.html [Accessed: May 8, 2018]

[32] Kaufman JC, Beghetto RA. Beyond big and little: The four c model of creativity. Review of General Psychology. 2009;**13**:1-12. Available from: https://pdfs.semanticscholar.org/f4fc/c5125a4eb702cdff0af421e500433fbe9a16.pdf

[33] Papert SA. Mindstorms: Children, Computers and Powerful Ideas. The Perseus Book Group; 1993

[34] Gilday D. How to build MindCub3r for LEGO MINDSTORMS EV3 [Online]. 2013. Available from: http://mindcuber.com/mindcub3r/mindcub3r.html [Accessed: May 8, 2018]

[35] PV Productions. What's a LEGO GBC? [Online]. 2018. Available from: https://pv-productions.com/whats-a-lego-gbc/ [Accessed: May 8, 2018]

[36] European Commission. European Business Forum on Vocational Training—Report from Workshop 3: The Use of Work Based Learning for Competence Development—Challenges and Practices [Online]. 2012. Available from: http://ec.europa.eu/education/vocational-education/doc/forum12/sumw3_en.pdf [Accessed: Dec 19, 2013]

[37] The Design Council. The 'Double Diamond' Design Process Model [Online]. 2008. Available from: http://webarchive.nationalarchives.gov.uk/20080821071133http://www.designcouncil.org.uk/en/About-Design/managingdesign/The-Study-of-the-Design-Process/ [Accessed: May 8, 2018]

[38] Pink DH. Drive: The Surprising Truth about What Motivates Us. Riverhead Books; 2009

[39] Rebrickable. Rebrickable [Online]. 2018. Available from: https://rebrickable.com [Accessed: May 8, 2018]

[40] PrintABrick. PrintABrick [Online]. 2017. Available from: https://printabrick.org/ [Accessed: May 8, 2018]